T0184284

Lecture Notes in Computer Science 12687

More information about this subseries at http://www.springer.com/series/7407

Rudolf Freund · Tseren-Onolt Ishdorj ·
Grzegorz Rozenberg · Arto Salomaa ·
Claudio Zandron (Eds.)

Membrane Computing

21st International Conference, CMC 2020
Virtual Event, September 14–18, 2020
Revised Selected Papers

Editors
Rudolf Freund
TU Wien
Vienna, Austria

Grzegorz Rozenberg
Leiden University
Leiden, The Netherlands

Claudio Zandron
University of Milano-Bicocca
Milan, Italy

Tseren-Onolt Ishdorj
Mongolian University of Science and Technology
Ulaanbaatar, Mongolia

Arto Salomaa
Turku Center for Computer Science
Turku, Finland

ISSN 0302-9743 ISSN 1611-3349 (electronic)
Lecture Notes in Computer Science
ISBN 978-3-030-77101-0 ISBN 978-3-030-77102-7 (eBook)
https://doi.org/10.1007/978-3-030-77102-7

LNCS Sublibrary: SL1 – Theoretical Computer Science and General Issues

This Springer imprint is published by the registered company Springer Nature Switzerland AG
The registered company address is: Gewerbestrasse 11, 6330 Cham, Switzerland

Preface

The present volume contains the invited contributions and a selection of papers presented at the 21st International Conference on Membrane Computing (ICMC 2020), which was held during September 14–18, 2020. Due to the pandemic situation related to the COVID-19 virus, it was organized as an electronic conference, joining together the 9th ACMC 2020, Asian Conference on Membrane Computing (ACMC 2020), originally planned to be held in Ulaanbaatar, Mongolia, and the 21st CMC 2020, International Conference on Membrane Computing (CMC 2020), originally planned to be held in Vienna, Austria. Further information can be found on the website at the following address: http://2020.e-icmc.org/.

The CMC series started with three workshops organized in Curtea de Argeş, Romania, in 2000, 2001, and 2002. The workshops were then held in Tarragona, Spain (2003), Milan, Italy (2004), Vienna, Austria (2005), Leiden, Netherlands (2006), Thessaloniki, Greece (2007), and Edinburgh, UK (2008).

The tenth edition was organized again in Curtea de Argeş, Romania, in 2009, where it was decided to continue the series as the Conference on Membrane Computing (CMC). The following editions were held in Jena, Germany (2010), Fontainebleau, France (2011), Budapest, Hungary (2012), Chişinău, Moldova (2013), Prague, Czech Republic (2014), Valencia, Spain (2015), Milan, Italy (2016), Bradford, UK (2017), and Jena, Germany (2018).

To celebrate the 20th edition, the conference was organized once again in Curtea de Argeş, Romania, in 2019.

A regional version of CMC, the Asian Conference on Membrane Computing, ACMC, started in 2012 in Wuhan, China, and continued in Chengdu, China (2013), Coimbatore, India (2014), Hefei, China (2015), Bangi, Malaysia (2016), Chengdu, China (2017), Auckland, New Zealand (2018), and Xiamen, China (2019).

The invited lectures were given by Henry N. Adorna (Department of Computer Science, UP Diliman, Philippines), Artiom Alhazov (Vladimir Andrunachievici Institute of Mathematics and Computer Science, Moldova), Lucie Ciencialova (Silesian University in Opava, Czech Republic), Gheorghe Păun (The Romanian Academy, Romania), Mario J. Pérez--Jiménez (Universidad de Sevilla, Spain), and Fan-Gang Tseng (National Tsing-Hua University, Taiwan).

The editors express their gratitude to the Program Committee, the invited speakers, the authors of the papers, the reviewers, and all the participants for their contributions to the success of ICMC 2020.

The support of the University of Vienna and of the Mongolian University of Science and Technology are gratefully acknowledged.

March 2021

Rudolf Freund
Tseren-Onolt Ishdorj
Grzegorz Rozenberg
Arto Salomaa
Claudio Zandron

Organization

Steering Committee of CMC and ACMC

Henry Adorna	University of the Philippines Diliman, Philippines
Artiom Alhazov	Vladimir Andrunachievici Institute of Mathematics and Computer Science, Moldova
Bogdan Aman	University of Iaşi, Romania
Matteo Cavaliere	Manchester Metropolitan University, UK
Erzsébet Csuhaj-Varjú	Eötvös Lorand University, Hungary
Giuditta Franco	Universitá di Verona, Italy
Rudolf Freund	Technische Universitat Wien, Austria
Marian Gheorghe (Honorary Member)	University of Bradford, UK
Thomas Hinze	Friedrich Schiller University of Jena, Germany
Florentin Ipate	University of Bucharest, Romania
Shankara N. Krishna	Indian Institute of Technology, India
Alberto Leporati	Universitá di Milano-Bicocca, Italy
Taishin Y. Nishida	Toyama Prefectural University, Japan
Linqiang Pan (Co-chair)	Huazhong University of Science and Technology, China
Gheorghe Păun (Honorary Member)	The Romanian Academy and Academia Europaea, Romania
Mario J. Pérez-Jiménez	Universidad de Sevilla, Spain
Agustín Riscos-Núñez	Universidad de Sevilla, Spain
Jose M. Sempere	Universidad de Valencia, Spain
Petr Sosík	Silesian University in Opava, Czech Republic
Kumbakonam Govindarajan Subramanian	University Sains Malaysia, Malaysia
György Vaszil	University of Debrecen, Hungary
Sergey Verlan	University Paris Est Créteil, France
Claudio Zandron (Co-chair)	Universitá di Milano-Bicocca, Italy
Gexiang Zhang	Chengdu University of Information Technology, China

Organizing Committee of ICMC 2020

Gordon Cichon	Mongolian University of Science and Technology, Mongolia
Rudolf Freund (Co-chair)	Technische Universitat Wien, Austria

Franziska Gusel	Technische Universitat Wien, Austria
Tseren-Onolt Ishdorj (Co-chair)	Mongolian University of Science and Technology, Mongolia
Sergiu Ivanov	Université d'Evry, France
Zoljargal Jargalsaikhan	Mongolian University of Science and Technology, Mongolia
Javzansuren Jigjidsuren	Mongolian University of Science and Technology, Mongolia
Chuluunbandi Naimannaran	Mongolian University of Science and Technology, Mongolia
Marion Oswald	Technische Universitat Wien, Austria
Namnan Tumur	Mongolian University of Science and Technology, Mongolia
Ewa Vesely	Technische Universitat Wien, Austria

Program Committee of ICMC 2020

Henry Adorna	University of the Philippines Diliman, Philippines
Artiom Alhazov	Vladimir Andrunachievici Institute of Mathematics and Computer Science, Moldova
Bogdan Aman	University of Iaşi, Romania
Altangerel Ayush	Mongolian University of Science and Technology, Mongolia
Catalin Buiu	University of Bucharest, Romania
Francis George C. Cabarle	University of the Philippines Diliman, Philippines
Matteo Cavaliere	Manchester Metropolitan University, UK
Lucie Ciencialova	Silesian University in Opava, Czech Republic
Erzsébet Csuhaj-Varjú	Eötvös Lorand University, Hungary
Rudolf Freund (Co-chair)	Technische Universitat Wien, Austria
Giuditta Franco	Universitá di Verona, Italy
Xiaoju Dong	Shanghai Jiao Tong University, China
Marian Gheorghe	Bradford University, UK
Ping Guo	Chongqing University, China
Juanjuan He	Wuhan University of Science and Technology, China
Thomas Hinze	Friedrich Schiller University of Jena, Germany
Florentin Ipate	University of Bucharest, Romania
Sergiu Ivanov (Co-chair)	Université d'Evry, France
Shankara N. Krishna	Indian Institute of Technology, India
Tseren-Onolt Ishdorj (Co-chair)	Mongolian University of Science and Technology, Mongolia
Savas Konur	Bradford University, UK
Alberto Leporati	Universitá di Milano-Bicocca, Italy
Jia Li	Chongqing University, China
Xiangrong Liu	Xiamen University, China
Xiyu Liu	Shangdon University, China

Ravie Chandren Muniyandi	Universiti Kebangsaan Malaysia, Malaysia
Ferrante Neri	University of Nottingham, UK
Radu Nicolescu	University of Auckland, New Zealand
Taishin Nishida	Toyama Prefectural University, Japan
David Orellana-Martín	Norwegian University of Science and Technology, Norway
Yunyun Niu	China University of Geosciences, China
Linqiang Pan (Co-chair)	Huazhong University of Science and Technology, China
Andrei Pun	University of Bucharest, Romania
Gheorghe Păun	The Romanian Academy and Academia Europaea, Romania
Hong Peng	Xihua University, China
Mario Pérez-Jiménez	Universidad de Sevilla, Spain
Agustín Riscos-Núñez	Universidad de Sevilla, Spain
Haina Rong	Southwest Jiaotong University, China
Jose M. Sempere	Universidad de Valencia, Spain
Bosheng Song	Hunan University, China
Tao Song	China University of Petroleum, China
Petr Sosík	Silesian University in Opava, Czech Republic
K. G. Subramanian	University Sains Malaysia, Malaysia
D. G. Thomas	Madras Christian College, India
György Vaszil	University of Debrecen, Hungary
Sergey Verlan	University Paris Est Créteil, France
Jun Wang	Xihua University, China
Tingfang Wu	Soochow University, China
Jianhua Xiao	Nankai University, China
Jie Xue	Shandong Normal University, China
Hsu-Chun Yen	National Taiwan University, R.O.C
Jianying Yuan	Xihua University, China
Claudio Zandron (Co-chair)	Universitá di Milano-Bicocca, Italy
Xiangxiang Zeng	Xiamen University, China
Gexiang Zhang (Co-chair)	Xihua University, China
Xingyi Zhang	Anhui University, China
Xue Zhang	Tufts University, USA
Xuncai Zhang	Zhengzhou University of Light Industry, China
Ming Zhu	Xihua University, China

Contents

Transition Graphs of Reversible Reaction Systems

Attila Bagossy and György Vaszil(✉)

Department of Computer Science, Faculty of Informatics, University of Debrecen,
Kassai út 26, Debrecen 4028, Hungary
{bagossy.attila,vaszil.gyorgy}@inf.unideb.hu

Abstract. We study the transition graphs, and thus, the possible computational paths of reaction systems which are reversible according to different notions of reversibility. We show that systems which are reversible in the sense of our earlier work produce very simple types of transition graphs. A somewhat more complicated, but still quite simple class of transition graphs is obtained if we consider so-called initialized reversible systems. Finally we introduce the notion of reversibility with lookbehind, and show that systems which are reversible in this sense produce the same transition graphs (and thus, the same computations) as the state transition diagrams of reversible finite transition systems.

1 Introduction

Reaction systems, introduced by Ehrenfeucht and Rozenberg in [6], aim to capture biochemical processes occurring inside living cells. The main intuition behind this model of computation is the interplay between facilitation and inhibition as these mechanisms define which reactions can take place, and thus how computations proceed. A reaction system is a set of reactions, each reaction is represented by a triple of finite sets: the reactants, the inhibitors, and the results. In each step, the system produces resulting elements according to the set of reactants and the set of reactions that are not inhibited. This core idea is further complemented by the model's two distinctive characteristics. In contrast to multiset-based frameworks, reaction systems present a qualitative approach in which if an element (or reactant) is present, then it is assumed to be available in the necessary amount. As a consequence of this principle, reactions may freely use the same resource and will not interfere with each other. The second characteristic is the concept of no permanency which means that if there is no reaction sustaining a particular element, then the element will vanish. Building on these principles, reaction systems perform computations in so-called interactive processes that combine the result of reactions with input from the enclosing environment.

The work of Gy. Vaszil was supported by the National Research, Development and Innovation Fund of Hungary through project no. K 120558, financed under the K 16 funding scheme.

R. Freund et al. (Eds.): CMC 2020, LNCS 12687, pp. 1–16, 2021.
https://doi.org/10.1007/978-3-030-77102-7_1

Since its inception, the model of reaction systems received vast research interest thanks to its unique properties and easy-to-extend nature. Research topics, for example, include the study of the state transition function defined by a particular system (see [5] among others), the introduction of time (see [11]), or modules (see [8]). For a more comprehensive enumeration, the reader is referred to [4,7].

In this paper we are going to study the transition graphs of reversible reaction systems. Transition graphs were introduced in [9] to represent the global dynamics of these systems. Such a graph is a directed graph where each vertex is a state of the system (represented as a set of elements present in the system at a given step of the computation) and directed edges from a vertex lead to the vertices representing the new states of the system which can be reached after all reactions enabled at the origin, with possible additions from the outside environment, are performed. The notion of reversibility was also studied in this framework. Reversible processes which do not depend on input from the external environment were considered in [1], and a somewhat more general notion of reversibility when certain kinds of inputs are allowed was proposed in [3].

Starting with our previously established definitions regarding reversible reaction systems in [3], we are going to study and compare the transition (or behavior) graphs corresponding to different reversible system definitions. Instead of examining the properties of individual interactive processes (that can be thought of as computational paths), we take a graph that describes every possible process in a given system and study how varying the underlying definitions (especially the different possible notions of reversibility) affects the graphs. By exploring these graphs, we infer the computational properties of the various definitions of reversible systems. We will consider the following variations.

1. Reversible systems as defined in our previous work [3]. We will consider two variants: First, we will use our notion of reversibility combined with the original definitions given in [6] where the initial state of the system is defined by the first environmental input, then we will use a slight modification which allows for arbitrary elements in the initial states of interactive processes.
2. Systems which are reversible with lookbehind. This is a modified notion of reversibility allowing interactive processes to examine not only the current result set but the previous environmental input as well.

The rest of the paper is organized as follows. In Sect. 2 we provide a brief introduction to the fundamental notions of reaction systems and the notion of reversibility as introduced in [3]. Then in Sect. 3 and Sect. 4 we present the transition graphs for the above mentioned systems as well as the comparison between them. Finally, Sect. 5 closes the paper with some conclusions.

2 Preliminaries

In this section, we first briefly introduce the essential concepts of reaction systems. We refer the reader to [4,6] for a more comprehensive description. Concerning reversibility, we only cover the most important definitions and requirements, see [3] for detailed results and proofs. For more information on transition graphs, refer to [11].

Reaction systems model biochemical reactions by the interplay of facilitation and inhibition. Reactions are defined over a finite set of entities (usually denoted by S) and every *reaction* a is a triplet of three finite sets $a = (R_a, I_a, P_a)$ (where each of these sets are subsets of S). The sets R_a, I_a and P_a contain the *reactants*, *inhibitors* and *products* of the reaction, respectively, and they satisfy the following constraints: First, the set of reactants and the set of inhibitors are disjoint ($R_a \cap I_a = \emptyset$), otherwise the reaction would never be applicable, as we will see later. Second, the set of reactants and the set of products are non-empty ($R_a \neq \emptyset$ and $P_a \neq \emptyset$). It is usually also assumed that the set of inhibitors is non-empty, but for the sake of being as general as possible, we will drop this additional assumption here. The set of all reactions over S is denoted by $\mathrm{rac}(S)$.

Remark 1. In what follows, if a is a reaction, then we will denote its components as R_a, I_a and P_a without explicitly writing out the complete triplet form $a = (R_a, I_a, P_a)$.

Based on the core idea of the model, a reaction is applicable (or enabled) if all of its reactants and none of its inhibitors are present. Applying a reaction creates its products. These intuitions are formalized as follows.

Given a set of arbitrary symbols (or entities) S and a reaction $a \in \mathrm{rac}(S)$, a is *enabled by* $W \subseteq S$ if $R_a \subseteq W$ and $I_a \cap W = \emptyset$. The *result of* a *on* W, denoted by $\mathrm{res}_a(W)$, is defined as $\mathrm{res}_a(W) = P_a$ if a is enabled by W, or $\mathrm{res}_a(W) = \emptyset$ if a is not enabled by W.

If A is a finite set of reactions over S, then $\mathrm{en}_A(W)$ denotes the *set of all reactions in* A *enabled by* W, thus $\mathrm{en}_A(W) = \{a \in A \mid a \text{ is enabled by } W\}$, and the *result of* A *on* W, denoted by $\mathrm{res}_A(W)$, is defined as $\mathrm{res}_A(W) = \bigcup_{a \in A} \mathrm{res}_a(W)$.

With these definitions in mind, we can now see how the characteristics mentioned in the Introduction are implemented. Reactions with overlapping reactant sets do not interfere as each one is allowed to create its products if enabled. This non-interefering nature also applies to the products. Even if the same entity is produced by multiple reactions, still there will be a single occurrence in the result set as reaction systems are defined over sets instead of multisets.

Prior to defining reaction systems and how they perform computation, we introduce further shorthand notations to ease our work with reactants and products of finite sets of reactions.

Notation 1. Let A be a finite set of reactions. Then, we denote by R_A and P_A the union of the reactant sets and product sets, respectively: $R_A = \cup_{a \in A} R_a$ and $P_A = \cup_{a \in A} P_a$.

If S is a finite set such that $A \subseteq \mathrm{rac}(S)$, then $\mathrm{EN}_A(S)$ contains the sets of reactions where the members of each set can be applied together for some subset of S. Formally

$$\mathrm{EN}_A(S) = \{E \subseteq A \mid \text{there exists } S' \subseteq S, \text{ such that } \mathrm{en}_A(S') = E\}.$$

Further, we denote by $\mathrm{RES}_A(S)$ the set that contains the results of applying every set of reactions in $\mathrm{EN}_A(S)$ to the appropriate subsets of entities, or formally

$$\mathrm{RES}_A(S) = \{\mathrm{res}_E(S') \mid S' \subseteq S, E \subseteq A, \text{ such that } \mathrm{en}_A(S') = E\}.$$

Example 1. Let us consider the set of reactions $A = \{a, b, c\}$ over $S = \{1, 2, 3\}$, where

$$a = (\{1\}, \emptyset, \{2\}),\ b = (\{2\}, \emptyset, \{3\}),\ c = (\{3\}, \{1\}, \{1\}).$$

Here, we have $\mathrm{EN}_A(S) = \{\{a\}, \{b\}, \{c\}, \{a, b\}, \{b, c\}\}$, since there is no set of elements such that a and c are enabled together (as $R_a \cap I_c \neq \emptyset$), that is, $\mathrm{en}_A(\{1\}) = \mathrm{en}_A(\{1, 3\}) = \{a\}$, $\mathrm{en}_A(\{1, 2\}) = \mathrm{en}_A(\{1, 2, 3\}) = \{a, b\}$, and we also have $\mathrm{en}_A(\{2\}) = \{b\}$, $\mathrm{en}_A(\{3\}) = \{c\}$, $\mathrm{en}_A(\{2, 3\}) = \{b, c\}$ in addition.

The elements of $\mathrm{RES}_A(S)$ are the product sets produced by the reactions in the sets of $\mathrm{EN}_A(S)$ applied to appropriate subsets of S

$$\mathrm{RES}_A(S) = \{\{2\}, \{3\}, \{1\}, \{2, 3\}, \{1, 3\}\}.$$

With the essential notions defined for reactions, we now recall the definition of a *reaction system*, which is an ordered pair $\mathscr{A} = (S, A)$. The background set S is a finite set of entities while $A \subseteq \mathrm{rac}(S)$ is the set of reactions.

Let $\mathscr{A} = (S, A)$ be a reaction system and let $n \geq 0$ be an integer. An *interactive process in* $\mathscr{A}$ is a pair $\pi = (\gamma, \delta)$ of finite sequences, such that

- γ is the *context sequence* of π, defined as $\gamma = C_0, C_1, \ldots C_n$, where $C_i \subseteq S$ for all $0 \leq i \leq n$, and
- δ is the *result sequence* of π, defined as $\delta = D_0, D_1, \ldots D_n$, where $D_0 = \emptyset$ and $D_i = \mathrm{res}_A(D_{i-1} \cup C_{i-1})$ for all $1 \leq i \leq n$.

We also define $\mathrm{sts}(\pi)$ as the *state sequence* of π by

- $\mathrm{sts}(\pi) = W_0, W_1, \ldots W_n$, where $W_i = C_i \cup D_i$ for all $0 \leq i \leq n$.

The above notion of an interactive process is visualized in Fig. 1. Note that the context and results sets are not required to be disjoint, although the figure shows them as non-overlapping rectangles.

As a consequence of this definition, every interactive process is finite with predetermined length. Therefore an interactive process can be thought of as a finite sequence of states (the union of contexts and results). The idea of input from the surrounding environment is formalized by the context sequence. A process in which every context set is empty is said to be *context-independent.*

Interactive processes also encompass the concept of no permanency. Every new state consists of the products of the previous state and the environmental input. Hence, if an entity is not produced by any enabled reaction and is not present in the context set, then it will disappear. This idea stems from abstract biochemistry where an entity must be sustained by some active process. In the absence of such a process, the entity will vanish. See [4,6] for motivations and more details.

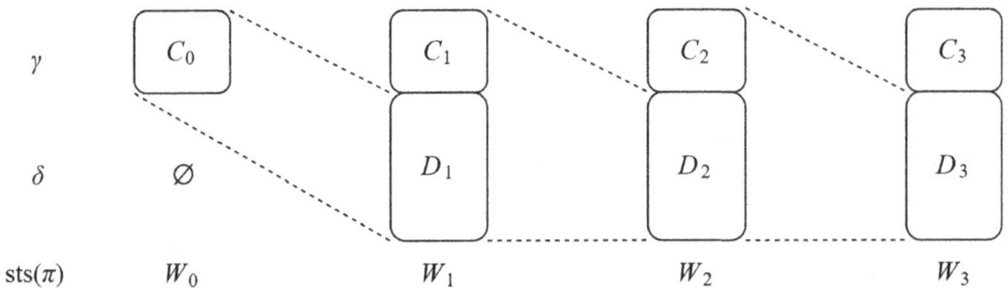

Fig. 1. An interactive process π in a reaction system.

Now we present a possible notion of reversibility for reaction systems which we introduced and investigated in [3]. Generally, a sequential model of computation (such as the model of reaction systems) might be considered reversible if it is "backward deterministic", that is, if no "configuration" is accessible from two different configurations (in a way which is "undistinguishable"), or in other words, every configuration has a *unique predecessor*. The point is, that given any "state" of the system, we should be able to determine the preceding computational step. Reaction systems perform computations via interactive processes, thus we should interpret the concept of configuration (or "state") and the concept of unique predecessor for such interactive processes.

In [3] we have followed the natural idea to identify the configurations we are interested in (from the point of view of reversibility) with the states of interactive processes as defined above to be the union of the result sets and the corresponding context sets.

Definition 1. ([3]) *Let* $\mathscr{A} = (S, A)$ *be a reaction system and* $\pi = (\gamma, \delta)$ *be an interactive process in* $\mathscr{A}$*, such that* $\gamma = C_0, C_1, \ldots C_n$ *and* $\mathrm{sts}(\pi) = W_0, W_1, \ldots W_n$.

A state W_i*,* $1 \leq i \leq n$*, has* multiple predecessors *if there exists* $W \subseteq S$ *such that* $W \neq W_{i-1}$*, but* $\mathrm{res}_A(W) \cup C_i = W_i$*. If there is no such* W*, then* W_i *has a* unique predecessor.

The interactive process π *is* reversible *if every state* W_i*,* $1 \leq i \leq n$*, has a unique predecessor.*

After defining reversibility for individual processes, we would like to continue by defining reversible reaction systems. As a first attempt, one might introduce reversible reaction systems as ones with reversible interactive processes only. This definition, however, needs some further refinement. To see this, consider the following.

Since reaction systems are finite both in terms of reactions and entities, regardless of which state we begin with, we eventually run out of enabled reactions, or start to loop. In the former case, as there are no applicable reactions, the produced result set is the empty set. According to the definition of reversible interactive processes, no state is allowed to have multiple predecessors, therefore, in order to be reversible, we need to ensure that there is a unique combination of reactions producing the empty result set as well. This is greatly limiting, however, as loops not involving the empty result set are forbidden (since entering and continuing the loop corresponds to two distinct predecessors), so the reaction system may only perform a single, pre-determined computation from the initial state to the empty result set. Reversible interactive processes in such a case, could only contain a finite subsequence of states from this computation.

According to these considerations, it seems reasonable to not take into account every interactive process in a reaction system when defining reversibility. Thus, in what follows, we simply sidestep the issue of no applicable reactions by defining so-called non-restarting interactive processes.

Let $\mathcal{A}$ be a reaction system and $\pi = (\gamma, \delta)$ be an interactive process in $\mathcal{A}$ such that $\delta = D_0, D_1, \ldots D_n$. The interactive process π is *non-restarting* if $D_i \neq \emptyset$, $1 \leq i \leq n$. If the opposite holds, then π is *restarting.*

Definition 2. ([3]) *A reaction system $\mathcal{A}$ is* reversible, *if every non-restarting interactive process in $\mathcal{A}$ is reversible.*

Before we continue, we would like to present the requirements this definition poses on reversible interactive processes. As simple as this notion of a reversible process is, it has great consequences on the possible reactions and context entities of the enclosing system. For a formal description of the necessary and sufficient conditions reaction systems need to fulfill in order to be reversible, see Theorem 1 of [3]. Here we present the basic ideas without proofs in a less formal manner.

In the following discussion, we assume a finite background set S and a finite set of reactions $A \subseteq \text{rac}(S)$. In order for a reaction system to be reversible, it must fulfill the following conditions.

1. If $E_1, E_2 \subseteq A$ are different sets of reactions in $\text{EN}_A(S)$, they must produce different result sets, or in other terms, $E_1, E_2 \in \text{EN}_A(S)$ with $E_1 \neq E_2$ implies $P_{E_1} \neq P_{E_2}$. Clearly, if we had two different sets of reactions that produce the same entities when applied, then the state formed by these entities would have multiple predecessors.
2. If we take any two distinct subsets of S, then the reactions enabled by these sets must be different as well. In other words, if $T_1, T_2 \subseteq S$ with $T_1 \neq T_2$ and $\text{en}_A(T_1) \neq \emptyset$, then $\text{en}_A(T_1) \neq \text{en}_A(T_2)$ must also hold. To see this, consider that if two subsets would enable the same set of reactions, then the result set of these reactions would be a state having at least two predecessors (the very subsets T_1, T_2 we took from S).
3. Finally, consider the result sets of the reaction sets in $\text{EN}_A(S)$. If we are able to transform one such result set R_1 to another, say R_2, by augmenting it with entities which can also appear in the context (as part of a context set of some interactive process), then the state $W = R_2$ will have multiple predecessors. To see this, let $D = R_2$, $C = \emptyset$ and $D' = R_1$, $C' = R_2 \setminus R_1$, then consider the state $W = D \cup C = D' \cup C'$. If $R_2 = D = \text{res}_{E_2}(W')$ and $R_1 = D' = \text{res}_{E_1}(W'')$, then W has at least two predecessors, W' and W''. (We can obtain W from W' by applying the reactions in E_2 and adding the context C, or we can obtain W also from W'' by applying the reactions in E_1 and adding the

context set C'.) To formalize this condition, we "refine" the background set as $S = \Sigma_c \cup \Sigma_p$, the union of (not necessarily disjoint) alphabets of symbols where $\Sigma_c \subseteq S$ contains those entities which can appear as environmental input in the context sets of interactive processes, and $\Sigma_p \subseteq S$ containing those which can appear as products of reactions. (This is similar to so-called context-restricted reaction systems studied in [11].) Using this notation, we can formalize the above idea by requiring that having $R_1, R_2 \in \text{RES}_A(S)$ with $R_1 \neq R_2$ should also imply that $R_1 \setminus \Sigma_c \neq R_2 \setminus \Sigma_c$.

3 Transition Graphs of Reversible Systems

When introducing reversibility into a particular model of computation, the question naturally arises, how this affects the computational properties of the model. In the case of reaction systems, interactive processes are the only means of computation, thus when examining the higher-level computational properties of a specific system, we should start by looking at the contained processes. Since any system may only contain a finite number of possible states, so-called transition graphs offer a concise way of depicting every possible interactive process a particular system may enclose. In this section, we introduce the definition of transition graphs and then explore the graphs generated by the reversible systems of Sect. 2.

In what follows, we first introduce reachable result sets, that will eventually form the vertices of the transition graphs. Assuming the standard definition of interactive processes (see Sect. 2), D_0 must be empty, hence there might be result sets that cannot occur in any interactive process. By considering reachable result sets only, we exclude these sets from the transition graphs.

Definition 3. *Let $\mathscr{A} = (S, A)$ be a reversible reaction system with $S = \Sigma_p \cup \Sigma_c$ (where Σ_p and Σ_c are not necessarily disjoint). The result set $D \subseteq \Sigma_p$ is* reachable *if there exists a non-restarting interactive process $\pi = (\gamma, \delta)$ in $\mathscr{A}$ with $\delta = D_0, D_1, \ldots D_n$ such that $D = D_i$ for some $0 \leq i \leq n$. The set of reachable result sets in $\mathscr{A}$ is denoted by* $\text{REACH}_{\mathscr{A}}$.

If D_0 was allowed to be non-empty (thus, the reaction system can initiate its computation from an arbitrary result set), then every result set is reachable (since, we can freely choose the initial result set). By requiring D_0 to be empty, we restrict the possible result sets in interactive processes to the reachable sets of Definition 3. The set of reachable results sets is, in turn, determined by the reactions of the underlying reaction system.

Transition graphs were first introduced in [9] as vertices representing the subsets of the background set (usually denoted as S) connected by directed edges equivalent to the relation of "can be obtained from". Formally, the edge set is defined as $E = \{(W_1, W_2) \mid W_1 \subseteq S,\ \text{res}_A(W_1) \subseteq W_2\}$. Here, we would like to underline the subset relationship between the underlying sets of the connected vertices. This transition graph definition incorporates context sets (environmental input) by considering two states connected, if the result of the former can be augmented with context to form the latter.

As we are exclusively interested in subsets of S that appear in interactive processes, we will modify this notion to include only reachable results sets as vertices. Furthermore, since we wanted to put more emphasis on the role of the input, in our definition of transition graphs, edges are labeled with input sets from the environment. Such a labeled edge is drawn between two vertices if applying reactions to the union of the source vertex and the label produces the destination vertex as a result.

Definition 4. *Let $\mathscr{A} = (S, A)$ be reaction system, with $S = \Sigma_c \cup \Sigma_p$ as above. The* transition graph *of $\mathscr{A}$ is the graph $\mathrm{TG}_{\mathscr{A}} = (V, E)$, where $V = \mathrm{REACH}_{\mathscr{A}}$ is the set of vertices and*

$$E = \{\, (D, C, D') \mid D, D' \in V \text{ and } C \subseteq \Sigma_c \text{ such that } \mathrm{res}_A(D \cup C) = D' \,\}$$

is the set of directed edges with D being the starting vertex, D' the end vertex and C the label.

Example 2. Let $\mathscr{A}$ be a reaction system in which $\Sigma_p = \{1, 3, 5\}$ is the product alphabet, $\Sigma_c = \{0, 2, 4\}$ is the context alphabet and $A = \{a, b, c\}$ is the set of reactions, where

$$a = (\{0\},\ \{2, 4\},\ \{1\}),\ \ b = (\{1, 2\},\ \{0\},\ \{3\}),\ \ c = (\{1, 4\},\ \{0\},\ \{5\}).$$

Then, $\mathrm{TG}_{\mathscr{A}}$ consists of the vertices $V = \{\, \emptyset, \{1\}, \{3\}, \{5\}, \{3, 5\} \,\}$ and edges

$$E = \{\ (\emptyset, \{0\}, \{1\}), (\{1\}, \{2\}, \{3\}), (\{1\}, \{4\}, \{5\}), (\{1\}, \{2, 4\}, \{3, 5\})\ \}.$$

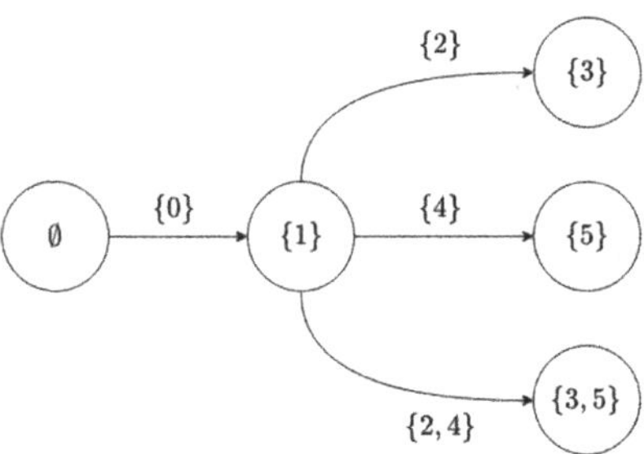

Fig. 2. Transition graph of the reaction system from Example 2.

With all the necessary notions in place, we now continue by examining the transition graphs of reversible reaction systems. In such systems, each result set may be the result of exactly one other state. As a consequence, for example, loops are forbidden (explained in more detail and proved in Theorem 1), which puts a firm constraint on the complexity of the non-restarting interactive processes. Therefore, the transition graphs of these systems are rather simple, they only contain finite computational branches (Fig. 2).

Theorem 1. *If $\mathscr{A}$ is a reversible reaction system, then the transition graph of $\mathscr{A}$ is either a single vertex or a directed rooted tree with all the edges pointing away from the root.*

Proof. Let $\mathscr{A} = (S, A)$ be a reversible reaction system with $S = \Sigma_p \cup \Sigma_c$ (where Σ_p and Σ_c are not necessarily disjoint), and let $\mathrm{TG}_{\mathscr{A}}$ be its transition graph.

By definition, $\mathrm{TG}_{\mathscr{A}}$ only includes an edge between two vertices if they are subsequent result sets in some non-restarting interactive process. Thus, the vertex for the empty result set does not have any incoming edges. Since the empty set is the initial result set (D_0) of every non-restarting interactive process in any reaction system, it will always be included in the transition graph.

If there is no context set C such that $\mathrm{en}_A(\emptyset \cup C) \neq \emptyset$ then $\mathrm{TG}_{\mathscr{A}}$ consists of a single vertex: the empty result set.

Now we show that $\mathrm{TG}_{\mathscr{A}}$ is a directed rooted tree if it has multiple vertices. A graph is a directed rooted tree if there is exactly one path between the root vertex and any other vertex. Since every vertex in the transition graph is a result set in some non-restarting interactive process, there must be a path between the root vertex and the vertex representing this set. Thus, we know that at least one path must exist connecting the root vertex with every other vertex.

Since $\mathscr{A}$ is reversible, every state has a unique predecessor. Consequently, every result set has a unique predecessor. As the vertices in the transition graph represent result sets and edges represent predecessor/successor relations, this means that every vertex other than the root has exactly one incoming edge. Therefore, there must be at most one path going from the root vertex to every other vertex. Because both the lower and the upper bound are equal to one, we have that there is a single path from the root vertex to any other vertex. Thus, if $\mathrm{TG}_{\mathscr{A}}$ has more than one vertex, it's a directed rooted tree with all edges pointing away from the root. □

As a consequence of the above theorem, non-restarting interactive processes in reversible systems (that are essentially computations) are just paths in a finite tree. Since this is a rather strict limitation, we might start experimenting with small relaxations in the underlying definitions to give rise to more complex graphs (those with vertices having in-degree greater than one or even containing cycles).

Such modification of particular interest is concerned with the definition of interactive processes. In the standard setting, D_0 (the initial result set) is empty for every interactive process. If the context sets can incorporate arbitrary entities from the background set, this does not pose any constraint on the initial state. On the other hand, in the case when the context and the product alphabets are different (as in the case of reversible systems), the product and the context alphabet can be disjoint, some results sets may not even be reachable at all. Similar ideas motivated the introduction of so-called initialized context-restricted reaction systems in [11] where non-empty initial product sets D_0 are also allowed. Now, let us examine how non-empty D_0 sets affect the transition graphs of reversible reaction systems. Following [11], we call our model *initialized* reversible reaction systems.

Theorem 2. *If $\mathscr{A}$ is an initialized reversible reaction system, then every component of the transition graph of $\mathscr{A}$ is either*

- *a single vertex,*
- *a directed rooted tree with edges pointing away from the root, or*
- *one directed cycle, such that each vertex of the cycle can also be the root of a tree with edges pointing away from the cycle.*

Proof. Let $\mathscr{A} = (S, A)$ be an initialized reversible reaction system with $S = \Sigma_p \cup \Sigma_c$ (where Σ_p and Σ_c are not necessarily disjoint) and with interactive processes that might start with a non-empty D_0 set.

Since our definition for the transition graph is the same as in Theorem 1, the reversibility of $\mathscr{A}$ results in a maximum of one for the in-degree of every vertex.

Now, let us consider the components of the transition graph. Given a result set $D \subseteq \Sigma_p$, if there is no $W \subseteq S$ such that $\mathrm{res}_A(W) = D$ (in any of the non-restarting interactive processes of $\mathscr{A}$), then the in-degree of the vertex corresponding to D is equal to 0. In this case, this vertex is either a component in itself or the root of a directed rooted tree. The former holds, if no result set can be reached from D in any of the non-restarting interactive processes (thus, the out-degree of the vertex is zero), while the latter is proved in the proof of Theorem 1.

With the first two cases (single vertex and tree) covered, let us examine components with exactly one cycle. We already know, that vertices with in-degree equal to zero either form single-vertex components or act as tree roots. Therefore, we only need to consider components in which the in-degree of every vertex is equal to one (as we previously proved that no vertex has in-degree greater than one). In this case, the component must include at least one cycle, otherwise there could be vertices with no incoming edges. However, while a component can include a single cycle when the in-degree of every vertex is one, multiple cycles are not possible. Single cycle components can take the form of a "branching ring", where the component includes a ring (or cycle) at its core and each member of this ring can additionally be the root of a tree branching out. On the other hand, multiple cycles can only be realized by including at least one vertex in each cycle with an edge coming from the cycle itself and an edge coming from a vertex outside of the appropriate cycle. As such configurations are forbidden for transition graphs of reversible systems, all remaining components must form a branching ring. □

Example 3. Let $\mathscr{A} = (S, A)$ be an initialized reversible reaction system with $S = \Sigma_p \cup \Sigma_c$ where $\Sigma_p = \{1, 3, 5, 7, 9, 11, 13, 15\}$ is the product alphabet, $\Sigma_c = \{0, 2, 4, 6, 8, 10, 12\}$ is the context alphabet and $A = \{a, b, c, d, e, f, g\}$ is the set of reactions, where

$$a = (\{0,1\}, \{6\}, \{3\}), \quad b = (\{2,3\}, \emptyset, \{5\}), \quad c = (\{4,5\}, \emptyset, \{1\}),$$
$$d = (\{1,6\}, \{0\}, \{7\}), \quad e = (\{7,8\}, \{10\}, \{9\}), \quad f = (\{7,10\}, \{8\}, \{11\}),$$
$$g = (\{12,13\}, \emptyset, \{15\}).$$

The transition graph $\mathrm{TG}_{\mathscr{A}}$ consists of three components: a branching ring, a tree, and a single vertex, as shown in Fig. 3.

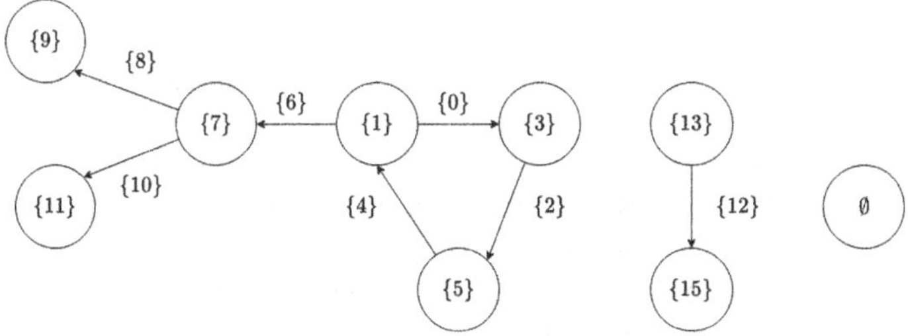

Fig. 3. Transition graph of the reaction system from Example 3.

As stated by Theorem 2, with a slight modification in the definition of interactive processes, we can achieve more involved transition graphs: those with multiple components and even a cycle per component. This allows for the occurrence of computations with loops, increasing the intuitive computational power of the model. Nevertheless, examined solely from an informal perspective, even with this change, we are unable to achieve the computational power of reversible finite automata, for example. The lack of higher in-degrees and multiple cycles places a severe constraint on the set of possible non-restarting interactive processes.

4 Reaction Systems Which Are Reversible with Lookbehind

As we have discussed in Sect. 2, we would like to look at the notion of reversibility as a kind of backward determinism, that is, given any "state" or configuration of the system, we should be able to determine the preceding computational step. In the previous section, following our earlier work in [3], we have interpreted the concept of configurations (or "states") as the state sets (that is, as the unions of product sets and context sets) of interactive processes. In this section we follow a different approach, one that is similar to how reversibility of (finite) automata is usually defined: The "state" of the machine is interpreted as the internal state of the finite control, together with additional information regarding the position of the reading head on the input tape, and the contents of the corresponding tape cell. See [2] and [12] for more details, or [10] for some more recent work regarding reversibility of finite automata. In such an interpretation, the predecessor configurations in a computation should be unique with respect to not only the internal state of the finite control, but also the input symbol that the reading head has just left behind, that is, the symbol that was read from the input tape in the previous computational step.

In this section we introduce the notion of reversibility with lookbehind by interpreting the "state" (or configuration) of reaction systems (in a similar manner as described above) as the current result set obtained in the previous computational step, together with the context set that was input in the same (that is, the previous) computational step. Thus, predecessor states of interactive processes should be unique with respect to the current product set, and the context set which was used to obtain the product, that is, the context set corresponding to the previous state of the interactive process.

In other words, our notion of reversibility for reaction systems which we recalled in Sect. 2 is based on the concept of unique predecessors. If a combination of reaction results and context entities (together forming what is called a state) has exactly one way to be produced, then it is said to have a unique predecessor. Hence, this definition focuses on the current result set and context set when determining which state occurred previously.

In this section, we introduce reversibility with lookbehind by taking a different approach to the definition of unique predecessors. Inspired by finite state automata, which are considered backward deterministic (and thus, reversible) if the current internal state and the previously consumed input symbol uniquely determine the previous internal state, reaction systems which are reversible with lookbehind can inspect both the current state and the previous context set. As a consequence, multiple state sets can produce the very same result sets (without losing the property of being reversible) given they include distinct context sets.

Definition 5. *Let* $\mathcal{A} = (S, A)$ *be a reaction system and* $\pi = (\gamma, \delta)$ *be an interactive process in* $\mathcal{A}$*, such that* $\gamma = C_0, C_1, \ldots C_n$*,* $\delta = D_0, D_1, \ldots D_n$ *and* $\text{sts}(\pi) = W_0, W_1, \ldots W_n$.

A state W_i*,* $1 \leq i \leq n$*, has* multiple predecessors with lookbehind *if there exist* $D \subseteq S$ *such that* $D \neq D_{i-1}$*, but* $\text{res}_A(D \cup C_{i-1}) = D_i$*. If there is no such* D*, then* W_i *has a* unique predecessor with lookbehind.

The interactive process π *is* reversible with lookbehind *if every state* W_i*,* $1 \leq i \leq n$*, has a unique predecessor with lookbehind.*

Now, we can define reversible systems using the above definition.

Definition 6. *A reaction system* $\mathcal{A}$ *is* reversible with lookbehind *if every non-restarting interactive process in* $\mathcal{A}$ *is reversible with lookbehind.*

Regarding transition graphs, an immediate consequence of the new definition of reversibility is the possibility of in-degrees higher than one. As shown in Fig. 4, despite the two incoming edges of the vertex $\{4\}$, when reversing the previous computation, we can now decide which one to take based on the preceding context (the label of the edge).

Continuing our previous discussion, we now compare the state diagrams and the transition graphs of reversible finite transition systems and reversible reaction systems with lookbehind, respectively. A finite transition system is usually denoted as a triplet $F = (Q, \Sigma, \delta)$, where Q is the set of states, Σ is the input alphabet and δ is the state transition function, mapping the current state and

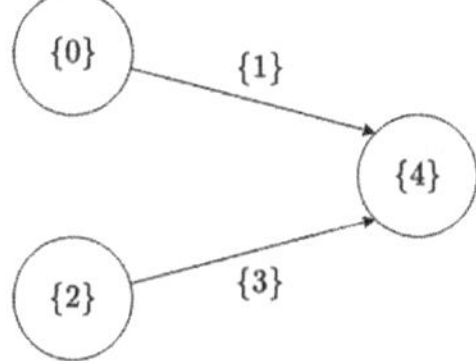

Fig. 4. Transition graph configuration that is not permitted for ordinary reversible systems, but is allowed in the case of systems which are reversible with lookbehind.

an input symbol to a result state. Finite transition systems differ from finite state automata by not having distinguished start and final states. Although the following results can be stated for finite state automata as well, reaction systems seem to be more closely related to transition systems, since interactive processes in reaction systems lack the concept of final (or accepting) state. This is further supported by [4], where a method is presented to convert finite transitions systems to reaction systems.

In what follows, when constructing transition graphs, we assume the definition of interactive processes in which D_0 (or, the initial result set) is empty. This means that every interactive process must start from the empty result set which makes it easier to create reaction systems from transition systems, as it allows for more control over the reachable result sets. Example 4 explores this idea in greater detail.

Example 4. Let F be a reversible finite transition system for which we want to construct a corresponding reaction system. Starting with the background set, we can create an entity for each input symbol of F as well as for each state of F. The entities created from the input symbols comprise the input alphabet Σ_c, while the entities representing the states belong to the product alphabet Σ_p. Now, we have that the result sets of the reaction system (D_i) represent the actual state of the underlying transition system, and the context sets correspond to the received input.

If the initial result set D_0 was allowed to be non-empty, then any entity of the product alphabet could be present in this set, even multiple entities. Since each entity represents a distinct state of the underlying transition system, the presence of multiple entities would mean that the transition system is in multiple states at once. As this is not permitted, one should require D_0 to be empty, since that way, the facilitation and inhibition aspects of the reaction can prevent such cases.

Theorem 3. *For every reversible finite transition system, there is a reaction system which is reversible with lookbehind and has the same states and transitions (apart from a starting state and its corresponding transitions).*

Proof. Let $F = (Q, \Sigma, \delta)$ be a reversible finite transition system. Then, the state diagram of F, $\mathrm{SD}_F = (V_F, E_F)$ is a directed graph, where $V_F = Q$ is the set of vertices and $E_F = \{(v, l, w) \mid \delta(v, l) = w\}$ is the set of directed, labeled edges.

Now, let us construct a reaction system $\mathscr{A} = (S, A)$ from F. Initially, we choose the background set S to be the union of two disjoint sets ($S = \Sigma_p \cup \Sigma_c$): the product alphabet corresponds to the states of the transition system (thus, $\Sigma_p = Q$), while the input alphabet is equivalent to the input alphabet of F (thus, $\Sigma_c = \Sigma$). We can define a set of reactions using the transition function of F:

$$\{(\{q, i\}, \Sigma_c \setminus \{i\}, \{r\}) \mid \delta(q, i) = r, \text{ for } q, r \in Q, \ i \in \Sigma\}.$$

Finite trasition systems do not have a designated initial state, but may begin their computation in an arbitrary state. In the case of reaction systems, however, an interactive process must start with the empty result set ($D_0 = \emptyset$). Since the context and the product alphabets are, in this case, disjoint ($\Sigma_p \cap \Sigma_c = \emptyset$), it is not possible to put a symbol representing some state $q \in Q$ in the initial context set (C_0). To overcome this issue and allow the reaction system to start its computation by jumping to an arbitrary state q, we introduce a new entity, α_q for each state $q \in Q$ and an appropriate reaction that lead from the empty result set to the result set representing this state q of F.

With this in mind, let us redefine the background and the reaction set of $\mathscr{A}$. The background set S is now the union of the following two sets: $\Sigma_p = Q$ and $\Sigma_c = \Sigma \cup \{\alpha_q \mid q \in Q\}$. The reaction set A is then defined as follows:

$$\begin{aligned} A =& \{(\{q, i\}, \Sigma_c \setminus \{i\}, \{r\}) \mid \delta(q, i) = r \text{ for } q, r \in Q, \ i \in \Sigma\} \cup \\ & \{(\{\alpha_q\}, \Sigma_p \cup (\Sigma_c \setminus \{\alpha_q\}), \{q\}) \mid q \in Q\}. \end{aligned}$$

The transition graph of $\mathscr{A}$ is defined as a directed graph based on the result sets and inputs of the non-restarting interactive processes in the system. Because of the definition of the transition graph, given any vertex q in the state diagram of F, we have a vertex in $\mathrm{TG}_{\mathscr{A}}$ corresponding to $D = \{q\}$. Additionally, for every edge (v, l, w) in the state diagram of F, we have an appropriate edge pointing from the vertex $D_1 = \{v\}$ to the vertex $D_2 = \mathrm{res}_{\mathscr{A}}(\{v, l\}) = \{w\}$ (because of the definition of A).

Consequently, apart from the vertex representing the empty state and its outgoing edges, the transition graph of $\mathscr{A}$ and the state diagram of F are isomorphic.

What is left to prove is that the states in the non-restarting interactive processes of $\mathscr{A}$ have unique predecessors with lookbehind (making $\mathscr{A}$ reversible with lookbehind). Since F is a reversible transition system, there is no vertex in its state diagram that has more than one incoming edge with the same label. As a consequence, each vertex in the transition graph of $\mathscr{A}$ satisfies the same property. Combining this fact with the disjointness of the product and input alphabets, we have that no state can be reached with the same input (or label, in the transition graph) from two different result sets. This is precisely the definition of having a unique predecessor with lookbehind. □

Theorem 4. *For every reaction system which is reversible with lookbehind, there is a reversible finite transition system with the same states and transitions.*

Proof. Let $\mathcal{A} = (S, A)$ be a reversible lookbehind reaction system with $S = \Sigma_p \cup \Sigma_c$ (where Σ_p and Σ_c are not necessarily disjoint). Also, let $\text{TG}_{\mathcal{A}} = (V_{\mathcal{A}}, E_{\mathcal{A}})$ be the transition graph of $\mathcal{A}$.

Now, let us construct a finite transition system $F = (Q, \Sigma, \delta)$ from $\mathcal{A}$. Using the transition graph of $\mathcal{A}$, we can define the states of the transition system as $Q = V_{\mathcal{A}}$. The input alphabet of F is going to contain the subsets of the input alphabet of $\mathcal{A}$, thus $\Sigma = 2^{\Sigma_c}$. By considering the edges in $\text{TG}_{\mathcal{A}}$, we can define the transition function as

$$\delta(q, i) = r, \text{ if there is an edge in } E_{\mathcal{A}} \text{ from } q \text{ to } r \text{ with label } i.$$

Because of the above definition of F, given any vertex (representing a result set) in the transition graph of $\mathcal{A}$, we will have a corresponding vertex in the state diagram of F. Furthermore, since the edges in the state diagram correspond to the transition function δ, which in turn was defined via the edges of $\text{TG}_{\mathcal{A}}$, we have that each edge in the state diagram of F will map to an edge in $\text{TG}_{\mathcal{A}}$. Thus, we have that the state diagram of F and the transition graph of $\mathcal{A}$ are isomorphic.

Analogous to the proof of Theorem 3, now we need to show that F is a reversible transition system. Since $\mathcal{A}$ is reversible, no vertex in its transition graph has more than one incoming edge with the same label. As the state diagram of F is isomorphic to $\text{TG}_{\mathcal{A}}$ and each edge has the same label as its counterpart in $\text{TG}_{\mathcal{A}}$, the same is true for each vertex in the state diagram. Thus, because the edges represent the state transitions induced by δ, we have that F is reversible. □

Based on the previous two theorems, we can state the following.

Proposition 1. *The state transition graphs of reversible finite transition systems correspond to the transition graphs of reaction systems which are reversible with lookbehind, apart from the special initial vertex corresponding to the initial empty result set of the reaction system. Vice-versa, the transition graph of any reaction system which is reversible with lookbehind corresponds to the state transition graph of a reversible finite transition system.*

5 Conclusion

In this paper, we have studied the possible computations of reversible reaction systems by examining their transition graphs. First, we considered reaction systems which are reversible according to our definition of reversibility given in [3] and concluded that the computational graphs (and so the possible computations) are very simple. Then we examined the same notion of reversibility for initialized reaction systems (see [11]) and obtained transition graphs which are somewhat more complicated, but still quite simple. Finally, motivated by the reversibility of (finite) automata, we introduced the notion of reversibility with lookbehind, which finally is able to produce functioning corresponding to the same types of transition graphs (and thus, the same possible computations) as the state transition diagrams of reversible finite transition systems.

To study other aspects of reaction systems which are reversible with lookbehind is a research topic that we would like to investigate in more detail in the future.

References

1. Aman, B., Ciobanu, G.: Controlled reversibility in reaction systems. In: Gheorghe, M., Rozenberg, G., Salomaa, A., Zandron, C. (eds.) CMC 2017. LNCS, vol. 10725, pp. 40–53. Springer, Cham (2018). https://doi.org/10.1007/978-3-319-73359-3_3
2. Angluin, D.: Inference of reversible languages. J. ACM **29**(3), 741–765 (1982)
3. Bagossy, A., Vaszil, G.: Simulating reversible computation with reaction systems. J. Membr. Comput. **2**(3), 179–193 (2020). https://doi.org/10.1007/s41965-020-00049-9
4. Brijder, R., Ehrenfeucht, A., Main, M., Rozenberg, G.: A tour of reaction systems. Int. J. Found. Comput. Sci. **22**, 1499–1517 (2011)
5. Dennunzio, A., Formenti, E., Manzoni, L., Porreca, A.E.: Reachability in resource-bounded reaction systems. In: Dediu, A.-H., Janoušek, J., Martín-Vide, C., Truthe, B. (eds.) LATA 2016. LNCS, vol. 9618, pp. 592–602. Springer, Cham (2016). https://doi.org/10.1007/978-3-319-30000-9_45
6. Ehrenfeucht, A., Rozenberg, G.: Reaction systems. Fundam. Inf. **75**(1–4), 263–280 (2007)
7. Ehrenfeucht, A., Kleijn, J., Koutny, M., Rozenberg, G.: Minimal reaction systems. In: Priami, C., Petre, I., de Vink, E. (eds.) Transactions on Computational Systems Biology XIV. LNCS, vol. 7625, pp. 102–122. Springer, Heidelberg (2012). https://doi.org/10.1007/978-3-642-35524-0_5
8. Ehrenfeucht, A., Rozenberg, G.: Events and modules in reaction systems. Theor. Comput. Sci. **376**, 3–16 (2007)
9. Genova, D., Hoogeboom, H.J., Jonoska, N.: A graph isomorphism condition and equivalence of reaction systems. Theor. Comput. Sci. **701**, 109–119 (2017)
10. Holzer, M., Kutrib, M.: Reversible nondeterministic finite automata. In: Phillips, I., Rahaman, H. (eds.) RC 2017. LNCS, vol. 10301, pp. 35–51. Springer, Cham (2017). https://doi.org/10.1007/978-3-319-59936-6_3
11. Męski, A., Penczek, W., Rozenberg, G.: Model checking temporal properties of reaction systems. Inf. Sci. **313**, 22–42 (2015)
12. Pin, J.-E.: On reversible automata. In: Simon, I. (ed.) LATIN 1992. LNCS, vol. 583, pp. 401–416. Springer, Heidelberg (1992). https://doi.org/10.1007/BFb0023844. https://hal.archives-ouvertes.fr/hal-00019977

Communicating Reaction Systems with Direct Communication

Erzsébet Csuhaj-Varjú[(✉)] and Pramod Kumar Sethy

Department of Algorithms and Their Applications, Faculty of Informatics,
Eötvös Loránd University ELTE, Budapest, Hungary
{csuhaj,pksethy}@inf.elte.hu

Abstract. We introduce and examine two variants of networks of reaction systems, called communicating reaction systems with direct communication, where the reaction systems send products or reactions to each other. We show that these types of networks of reaction systems can be obtained by simple mappings from single reaction systems. We also discuss some aspects of communication within these networks, and suggest open problems for future research.

1 Introduction

The theory of reaction systems has been a vivid research area recently. The concept of a reaction system was introduced by A. Ehrenfeucht and G. Rozenberg as a formal model of interactions between biochemical reactions. The interested reader is referred to [8] for the original motivation. The main idea of the authors was to model the behavior of biological systems in which a large number of individual reactions interact with each other.

A reaction system consists of a finite set of objects that represent chemicals and a finite set of triplets that represent chemical reactions. Each reaction consists of three nonempty finite sets: the set of reactants, the set of inhibitors, and the set of products. The set of reactants and the set of inhibitors are disjoint. Let T be a set of reactants. A reaction is enabled for T and it can be performed if all of its reactants are present in T and none of its inhibitors is in T. When the reaction is performed, then the set of its reactants is replaced by the set of its products. All enabled reactions are applied in parallel. The final set of products is the union of all sets of products that were obtained by the reactions that were enabled for T. For further details on reaction systems consult [9].

Reaction systems (R systems) are qualitative models, opposed to P systems (membrane systems) that are quantitative ones. The model of reaction systems focuses only on the presence or absence of the chemical species, and does not consider their amounts. Multiple reactions that have common reactants do not interfere. All of the reactions that are enabled at a certain step are performed simultaneously. Another feature of reaction systems which makes them different from other bio-inspired computational models, as for example, P systems, is the lack of permanency: the state of the system consists of only products of those

R. Freund et al. (Eds.): CMC 2020, LNCS 12687, pp. 17–30, 2021.
https://doi.org/10.1007/978-3-030-77102-7_2

reactions that took place in the last step. Those reactants that were not involved in any reaction disappear from the system. This property is widely used in the theory.

R systems have been studied in detail over the last 16 years. One interesting topic of their study is the theory of networks of reaction systems [4]. Such a construct is a virtual graph with a reaction system in each node. The reaction systems are defined over the same background set and work in a synchronized manner, governed by the same clock. After performing the reactions enabled for the current set of reactants at a node, certain products from other nodes can be added to the set of products at the node. The nodes, thus the reaction systems interact with each other using distribution and communication protocols. The set of products of each reaction system in the network forms a part of the environment of the network. Important ideas and results on these constructs can be found in [3,4].

In this paper we introduce the concept of communicating R systems with two variants of direct communication (cdcR systems, for short). These constructs are particular variants of networks of reaction systems [4]. Such a system consists of a finite set of extended versions of reactions defined over the same background set. These extended reaction systems (the components of the cdcR system), in addition to performing standard reactions, communicate either products or reactions to certain predefined target components. In the case of product communication, the products are associated with targets, i.e. labels of components which the product is sent to. In the case of reaction communication, each reaction is associated with a set of targets, labels of components. In this case, after performing the reaction, it is communicated to the target component. We note that the sender component can also be the target component. In both cases, after performing the reactions and the communication, the system performs a new transition. Communication is direct in these systems since the target of the product or the reaction to be communicated is explicitly given together with the cdcR system. We prove that for every cdcR system using any of the two types of direct communication (product or reaction), a reaction system can be constructed which simulates, up to some simple mapping(s), the given cdcR system. That is, these reaction systems provide representations of cdcR systems. We also discuss communication within the network, define static and active communication links, graphs, and describe how to represent active communication links and graphs of the cdcR systems under operation. We also compare the two communication variants. Finally, we provide conclusions and suggestions for future research.

2 Preliminaries

For basic notions of formal language and computation theory, the reader is referred to [11].

The set of all strings over an alphabet V is denoted by V^*, the set of nonempty strings by V^+. The empty string is denoted by λ and $|w|$ denotes the length of string w. A language L is a subset of V^*.

We recall the notions concerning reaction systems; most of them are taken from [8,9]. Some notations slightly differ from the standard ones; these changes are for technical reasons.

Definition 1. *Let S be a finite nonempty set; S is called the background set. A reaction ρ over S is a triplet (R_ρ, I_ρ, P_ρ) where R_ρ, I_ρ, P_ρ are nonempty subsets of S such that $R_\rho \cap I_\rho = \emptyset$.*

Sets R_ρ, I_ρ, P_ρ are called the sets of reactants, inhibitors, and products of ρ, respectively.

For convenience, reaction ρ will be given in the form $\rho : (R_\rho, I_\rho, P_\rho)$ in the sequel.

We consider now the effect of a reaction in a specific state of a reaction system; states are finite sets of entities.

Definition 2. *A reaction system is an ordered pair $\mathcal{A} = (S, A)$, where S is a background set and A is a finite nonempty set of reactions over S.*

Thus, a reaction system $\mathcal{A}$ is simply a set of reactions. In specifying A, we also give its background set S.

Definition 3. *Let S be a background set, $T \subseteq S, \rho : (R_\rho, I_\rho, P_\rho)$ be a reaction over S, and let A be a finite set of reactions over S. Then*

1. *ρ is enabled for T if $R_\rho \subseteq T$ and $I_\rho \cap T = \emptyset$;*
2. *the result of applying ρ to T, denoted by $res_\rho(T)$, equals P_ρ if ρ is enabled for T and $\emptyset$ otherwise;*
3. *the result of applying A to T, denoted by $res_A(T)$, is $\bigcup_{\rho \in A} res_\rho(T)$.*

Thus, reaction ρ is enabled for T if T contains all of the reactants of ρ and none of its inhibitors. If ρ is enabled for T, then its product will be a subset of the successor set of reactants. For $T \subseteq S$, $en_A(T)$ denotes the set of reactions of A that are enabled for T. Notice that res_A defines a function on 2^S, called the result function.

Definition 4. *The state sequence of a reaction system $\mathcal{A}$ with initial state T is given by successive iterations of the result function:*

$$(res^n_{\mathcal{A}}(T))_{n \in N} = (T, res_{\mathcal{A}}(T), res^2_{\mathcal{A}}(T), ...).$$

Since the background set of a reaction system is finite, the state space is also finite; thus, every state sequence is either finite or ultimately periodic.

3 Communicating Reaction Systems with Direct Communication

We introduce the concept of communicating R systems (cdcR systems) with two variants of direct communication. The concept is strongly related to the notion of a network of R systems [4] and it has been inspired by several variants of bio-inspired networks of language generating devices [5–7]. A cdcR system consists of a finite number of components, each component is a finite set of extended variants of reactions. Every component is defined over the same background set. The components, in addition to performing standard reactions, communicate products or reactions, according to the used protocol, to certain predefined target components. The components of the cdcR system work in a synchronized manner, governed by the same clock. In the case of product communication, the products are associated with targets, i.e. with the label of the component which the product is sent to. In the case of reaction communication, each reaction is associated with a set of targets, labels of a component. In this case, after performing the reaction, it is sent to the target components. We note that the target component can also be the sender component. In both cases, after performing the reactions and the communication, the system performs a new transition, i.e. the procedure is repeated. The reader may easily see that the targets define direct communication between the components. We show that for every cdcR system using any of the two types of communication a standard R system can be constructed which provides a representation of the given cdcR system; the operation of the two systems correspond to each other.

3.1 Communication by Products

We first define the notion of a cdcR system communicating by products.

Definition 5. *A cdcR system communicating by products (a cdcR(p) system, for short), of degree n,* $n \geq 1$, *is an* $(n+1)$*-tuple* $\Delta = (S, A_1, \ldots, A_n)$, *where*

- S *is a finite nonempty set, the background set of* Δ;
- A_i, $1 \leq i \leq n$, *is the ith component of* Δ, *where*
  - A_i *is a finite nonempty set of extended reactions of type pc (pc-reactions, for short).*
  - *Each pc-reaction* ρ *of* A_i *is of the form* $\rho : (R_\rho, I_\rho, \Pi_\rho)$, *where* R_ρ *and* I_ρ *are nonempty subsets of* S, $R_\rho \cap I_\rho = \emptyset$, *and* $\Pi_\rho \subseteq P_\rho \times \{1, \ldots, n\}$ *is a nonempty set with* P_ρ *being a nonempty subset of* S. R_ρ, I_ρ, Π_ρ *are called the set of reactants, the set of inhibitors, and the set of products with targets. A pair* (b, j), $1 \leq j \leq n$ *in* Π_ρ *means that product* $b \in S$ *is communicated to component* A_j.

The term *pc*-reaction means that the reaction communicates products.

We extend notions and notations concerning reaction systems to cdcR(p) systems. If it is clear from the context, for singleton sets $\{\rho\}$ we use notation ρ.

A *pc*-reaction $\rho : (R_\rho, I_\rho, \Pi_\rho)$ is enabled for the set $U \subseteq S$ if $R_\rho \subseteq U$ and $I_\rho \cap U = \emptyset$ as in case of standard reaction systems; this fact is denoted by $en_\rho(U)$. Let $U \subseteq S$ be a set of reactants and let ρ be a *pc*-reaction at component A_i. Then we define $res_\rho(U) = \{b \mid (b, i) \in \Pi_\rho\}$ if $en_\rho(U)$ and $res_\rho(U) = \emptyset$ otherwise.

Let $\Delta = (S, A_1, \ldots, A_n)$ be a cdcR(p) system and let $U \subseteq S$. We define $res_{A_i}(U) = \{b \mid (b,i) \in \Pi_\rho, \rho \in A_i, en_\rho(U)\}$ if at least one *pc*-reaction in A_i is enabled for U and $res_{A_i}(U) = \emptyset$ otherwise.

cdcR(p) systems operate by transitions, i.e. by changing their states. A state of a cdcR(p) systems $\Delta = (S, A_1, \ldots, A_n)$ is an n-tuple $(D_1, \ldots, D_n)$ where $D_i \subseteq S$, $1 \leq i \leq n$; D_i is called the state of component A_i, $1 \leq i \leq n$. Notice that D_i can be the empty set.

A transition in Δ means that every component of the cdcR(p) system performs all of its enabled *pc*-reactions on the current set of reactants and then communicates the obtained products to their target components, indicated in the corresponding *pc*-reaction. It is important to note that the same object (product) can be communicated to a component from several components and by several *pc*-reactions.

The sequence of transitions starting with an initial state forms a state sequence of Δ. Notice that by the definition of the *pc*-reactions, for a given initial state there is only one state sequence of Δ, i.e. for a given initial state, the sequence of transitions is deterministic.

Definition 6. *Let $\Delta = (S, A_1, \ldots, A_n)$, $n \geq 1$, be a cdcR(p) system.*

The sequence $\bar{D}_0, \ldots, \bar{D}_j, \ldots$ is called the state sequence of Δ starting with initial state $\bar{D}_0$ if the following conditions are met:
For every $\bar{D}_j$, $j \geq 0$ where $\bar{D}_j = (D_{1,j} \ldots, D_{i,j}, \ldots, D_{n,j})$, $1 \leq i \leq n$ it holds that $\bar{D}_{j+1} = (D_{1,j+1} \ldots, D_{i,j+1}, \ldots, D_{n,j+1})$ with
$D_{i,j+1} = \cup_{1 \leq k \leq n} Com_{k \to i}(res_{A_k}(D_{k,j}))$ where $Com_{k \to i}(res_{A_k}(D_{k,j})) = \{b \mid (b,i) \in \Pi_\rho, \rho : (R_\rho, I_\rho, \Pi_\rho) \in en_{A_k}(D_{k,j})\}$.

Sequence $D_{i,0}, D_{i,1}, \ldots$ is said to be the state sequence of component A_i of Δ, $1 \leq i \leq n$.

Notice that the state sequence does not end if $res_{A_i}(D_{i,j})$ is the empty set, since products can be communicated to the component in some later step.

Let $\Delta = (S, A_1, \ldots, A_n)$, $n \geq 1$, be a cdcR(p) system and let $\bar{D}_0, \bar{D}_1 \ldots$ be the state sequence of Δ starting with $\bar{D}_0$. Then every pair $(\bar{D}_i, \bar{D}_{i+1})$, $i \geq 0$ is said to be a transition in Δ and is denoted by $\bar{D}_i \Longrightarrow \bar{D}_{i+1}$.

We give an example for cdcR(p) systems.

Example 1. Let $\Delta = (S, A_1, A_2, A_3)$ be a cdcR(p) system where $S = \{a, b, c, d\}$ and components A_1, A_2 and A_3 are defined as follows. Let

$$A_1 = \{\rho_1 : (\{a,b\}, \{d\}, \{(a,2)\}), \rho_2 : (\{b\}, \{d\}, \{(b,2)\})\},$$

$$A_2 = \{\rho_3 : (\{a,b\}, \{c\}, \{(c,3)\}), \rho_4 : (\{a\}, \{c\}, \{(a,3)\})\},$$

$$A_3 = \{\rho_5 : (\{a,c\}, \{b\}, \{(a,1)\}), \rho_6 : (\{a\}, \{d\}, \{(b,1)\})\}.$$

Let $\bar{D}_0$, the initial state of Δ be given as $\bar{D}_0 = (\{a,b\}, \{a,b\}, \{a,c\})$. Then component A_1 performs both of its *pc*-reactions, ρ_1 and ρ_2, and communicates products a and b to component A_2. Similarly, A_2 performs both of its *pc*-reactions, ρ_3 and ρ_4, and communicates products c and a to component A_3.

As in the previous two cases, A_3 also performs both of its *pc*-reactions, ρ_5 and ρ_6. It communicates products a and b to component A_1. Thus, the new state of Δ will be $\bar{D}_1 = (\{a,b\},\{a,b\},\{a,c\})$, the same as $\bar{D}_0$.

If we change *pc*-reaction ρ_3 to ρ_3', where $\rho_3' : (\{c\},\{a,b\},\{(c,3)\})$, then only *pc*-reaction ρ_4 is enabled on $\{a,b\}$. Thus, after performing ρ_4 only product a is communicated to A_3. Thus, the new state of Δ in this case will be $(\{a,b\},\{a,b\},\{a\})$.

Next we show that every cdcR(p) system can be represented by an R system which provides a simulation as well in the following sense: the state sequences of the components of the cdcR(p) system can be obtained by simple mappings from the state sequence of the R system.

Theorem 1. *Let $\Delta = (S, A_1, \ldots, A_n)$, $n \geq 1$, be a cdcR(p) system and let $\bar{D}_0 = (D_{1,0}, \ldots, D_{n,0})$ be initial state of Δ. We can give a reaction system $\mathcal{A} = (S', A')$, initial state W_0 of $\mathcal{A}$, and mappings $h_i : 2^{S'} \to 2^S$ such that for each i, $1 \leq i \leq n$, the state sequence $D_{i,0}, D_{i,1}, \ldots$ of component A_i of Δ is equal to the sequence $h_i(W_0), h_i(W_1), \ldots$, where $W_0, W_1, \ldots$ is the state sequence of $\mathcal{A}$ starting from initial state W_0.*

Proof. To prove the statement, we first define the components of $\mathcal{A}$. Let $S' = \{[x,i] \mid x \in S,\ 1 \leq i \leq n\}$ be the background set of $\mathcal{A}$. For every i, $1 \leq i \leq n$ let $S_i' = \{[x,i] \mid x \in S\}$.

For any *pc*-reaction $\rho : (R_\rho, I_\rho, \Pi_\rho)$ of component A_i, $1 \leq i \leq n$, we define reaction $\rho' : (R_{\rho'}, I_{\rho'}, P_{\rho'})$ of $\mathcal{A}$ as follows: $R_{\rho'} = \{[x,i] \mid x \in R_\rho\}$, $I_{\rho'} = \{[y,i] \mid x \in I_\rho\}$, $P_{\rho'} = \{[x,k] \mid (x,k) \in \Pi_\rho, 1 \leq k \leq n\}$. $\mathcal{A}$ has no more reactions. It can immediately be seen that every reactant $[x,i]$ of $\mathcal{A}$ represents a reactant x in S that can be found at component A_i, and reversely. Thus, Δ and $\mathcal{A}$ correspond to each other, since by definition any reaction $\rho' : (R_{\rho'}, I_{\rho'}, P_{\rho'})$ of $\mathcal{A}$ where each element of $R_{\rho'}, I_{\rho'}$ is of the form $[x,i]$ corresponds to a *pc*-reaction $\rho : (R_\rho, I_\rho, \Pi_\rho)$ of component A_i, and reversely.

Let $W_0 = \{[x,i] \mid x \in D_{i,0}, 1 \leq i \leq n\}$ be the initial state of $\mathcal{A}$. It is easy to see that elements of W_0 correspond to elements of the initial states of the components of Δ.

Let us define for i, $1 \leq i \leq n$, mapping $h_i : 2^{S'} \to 2^S$ as follows. Let $U \subseteq S'$. If $U \cap S_i' \neq \emptyset$, then let $h_i(U) = \{x \mid [x,i] \in U\}$, otherwise let $h_i(U) = \emptyset$.

We prove that the state sequence of component A_i of Δ starting from initial state $D_{i,0}$ corresponds to the state sequence of $\mathcal{A}$ starting from W_0. For $j = 0$ and for any fixed i, $i \in \{1, \ldots, n\}$, $D_{i,0} = h_i(W_0)$, thus the statement for $j = 0$ holds. Suppose now that the statement holds for l, where $l \geq 1$, i.e. $D_{i,l} = h_i(W_l)$. We show that $D_{i,l+1} = h_i(W_{l+1})$ holds as well. The set of reactants $D_{i,l+1}$ is the union of two sets of reactants $U_{i,l+1}$ and $V_{i,l+1}$. $U_{i,l+1}$ consists of all products that are obtained by all enabled reactions of A_i performed on $D_{i,l}$ and which products do not leave the component A_i, i.e. which should be communicated to A_i. $V_{i,l+1}$ consists of all products of all enabled reactions performed on some $D_{k,l}$ which products leave component A_k, $k \neq i$. (Notice that the two sets U_{l+1} and $V_{i,l+1}$ can have joint elements.) Since $D_{i,l+1} = \cup_{1 \leq k \leq n} Com_{k \to i}(res_{A_k}(D_{k,l}))$

where $Com_{k \to i}(res_{A_k}(D_{k,l})) = \{b \mid (b,i) \in P_\rho \times \{1,\ldots,n\}, \rho = (R_\rho, I_\rho, \Pi_\rho) \in en_{A_k}(D_{k,l})\}$ and each *pc*-reaction $\rho = (R_\rho, I_\rho, \Pi_\rho)$ of component A_i corresponds to exactly one reaction $\rho' : (R_{\rho'}, I_{\rho'}, P_{\rho'})$ of $\mathcal{A}$ and reversely, where $R_{\rho'} = \{[x,i] \mid x \in R_\rho\}$, $I_{\rho'} = \{[y,i] \mid x \in I_\rho\}$, $P_{\rho'} = \{[x,k] \mid (x,k) \in \Pi_\rho, 1 \leq k \leq n\}$, it can be seen that $D_{i,l+1} = h_i(W_{l+1})$ holds. This implies that the statement of the theorem holds.

In the sequel, we also call reaction system $\mathcal{A}$ the flattened reaction system of Δ or a flattened version of Δ. Notice that a cdcR(p) system is allowed to have only one component, thus the use of the term flattened version is justified.

Definition 7. *Let $\Delta = (S, A_1, \ldots, A_n)$, $n \geq 1$, be a cdcR(p) system. Let reaction system $\mathcal{A} = (S', A')$ be defined as follows. Let $S' = \{[x,i] \mid x \in S,\ 1 \leq i \leq n\}$ be the background set of $\mathcal{A}$. For any pc-reaction $\rho : (R_\rho, I_\rho, \Pi_\rho)$ of component A_i, we define reaction $\rho' : (R_{\rho'}, I_{\rho'}, P_{\rho'})$ of $\mathcal{A}$ with $R_{\rho'} = \{[x,i] \mid x \in R_\rho\}$, $I_{\rho'} = \{[y,i] \mid x \in I_\rho\}$, $P_{\rho'} = \{[x,k] \mid (x,k) \in \Pi_\rho, 1 \leq k \leq n\}$. No other reaction is in A'. Then $\mathcal{A}$ is called the flattened reaction system of Δ.*

Based on the proof of Theorem 1 some observations can be made. We present the next statement without proof, since it is a direct consequence of Theorem 1 and its proof.

Corollary 1. *Let $\Delta = (S, A_1, \ldots, A_n)$, $n \geq 1$, be a cdcR(p) system and let $\mathcal{A} = (S', A')$ be an R system given as in Theorem 1. Furthermore, let $\bar{D}_0$ be the initial state of Δ and let W_0 be the initial state of $\mathcal{A}$ given as in the proof of Theorem 1. Then, for $m \geq 0$, a reactant $b \in S$ occurs at component A_i in the mth element of the state sequence of Δ starting with initial state $\bar{D}_0$ if and only if reactant $[b,i] \in S'$ occurs in the mth element of state sequence of $\mathcal{A}$ starting with initial state W_0.*

In [10,12] the following problem was discussed: For a given reaction system $\mathcal{A} = (S, A)$, a reactant $a \in S$ and $m \geq 2$ the decision problem whether a appears at the mth step of at least one state sequence of $\mathcal{A}$ is called the occurrence problem. Note that any nonempty subset of S can be considered as initial state of $\mathcal{A}$, thus the reaction system may have more than one state sequences. For some fixed values of the parameter m, the occurrence problem was shown to be NP-complete [12] and when m is given as input it is a PSPACE-problem [10].

We can formulate the occurrence problem for cdcR(p) systems as well. For a given cdcR(p) system $\Delta = (S, A_1, \ldots, A_n)$, $n \geq 1$, the problem whether a reactant $a \in S$ occurs at some component A_i at the mth element of the state sequence of Δ starting with some initial state $\bar{D}_0$ is called the occurrence problem of cdcR(p) systems. By Theorem 1, Corollary 1 and because to any reaction system we can construct a cdcR(p) system with only one component, we may state that the occurrence problem of cdcR(p) systems for some fixed values of m is NP-complete and it is a PSPACE-problem when m is given as input.

Next we deal with the communication of products within the cdcR(p) system under operation.

Definition 8. *Let $\Delta = (S, A_1, \dots, A_n)$, $n \geq 1$, be a cdcR(p) system. The static communication graph of Δ is a directed graph $\Gamma = (V, E)$, where V is the set of vertices (nodes) labeled with A_j, $1 \leq j \leq n$, and the set of edges E is defined by $E \subseteq \bar{A} \times \bar{A}$, where $\bar{A} = \{A_1, \dots, A_n\}$ and $(A_i, A_j) \in E$ if and only if there is a pc-reaction $\rho : (R_\rho, I_\rho, \Pi_\rho)$ in A_i such that Π_ρ contains an element (b, j).*

That is, from node A_i there is a directed edge to node A_j if and only if component A_i of Δ has at least one *pc*-reaction that communicates at least one product to component A_j.

Definition 9. *Let $\Delta = (S, A_1, \dots, A_n)$, $n \geq 1$, be a cdcR(p) system. Let $\bar{D}_0$ be an initial state of Δ and let $tr_l : \bar{D}_l \Longrightarrow \bar{D}_{l+1}$, $l \geq 0$ be a transition in the state sequence $\sigma : \bar{D}_0, \bar{D}_1, \dots, \bar{D}_l, \dots$ of Δ. If under transition tr_l, at component A_i at least one reaction is performed that communicates at least one product to component A_j, $1 \leq i, j \leq n$, then we say that there is an active communication link from component A_i to component A_j under transition $tr_l : \bar{D}_l \Longrightarrow \bar{D}_{l+1}$ in state sequence σ.*

The active communication graph $\Gamma_{tr_l} = (V, E_{tr_l})$ of Δ under transition tr_l in σ is defined as follows: V is given as for Γ and E_{tr_l} consists of all edges (A_i, A_j), $1 \leq i, j \leq n$ in E such that there is an active communication link from component A_i to component A_j under transition $tr_l : \bar{D}_l \Longrightarrow \bar{D}_{l+1}$.

Notice that the active communication graph is associated to a transition. Thus, if $\sigma : \bar{D}_0, \bar{D}_1, \dots$ of Δ is the state sequence of Δ starting from initial state $\bar{D}_0$, then σ defines a sequence of graphs Γ_{tr_i}, $i \geq 1$, where Γ_{tr_i} is the active communication graph associated to transition tr_i, $tr_i : \bar{D}_{i-1} \Longrightarrow \bar{D}_i$.

In the following we provide a representation of communication graphs (static and active) of cdcR(p) systems. In the proof of Theorem 1, we assigned to each product b of cdcR(p) system Δ a location, i.e. the number (label) of the component where the reactant is currently located. Thus, we used products of the form $[b, i]$ instead of b. This idea is extended in the following manner. In addition to the current place, the symbol describing the product will also code its previous location, the component from which it was communicated to its recent location. Thus, we will use symbols of the form $[b, i, j]$ meaning that a product b from component A_i is/was sent to component A_j. Using this variant of flattening the cdcR(p) system, we find a method for tracking active communication links associated to transitions in every given state sequence in Δ.

Theorem 2. *Let $\Delta = (S, A_1, \dots, A_n)$, $n \geq 1$ be a cdcR(p) system and let $\bar{D}_0$ be initial state of Δ. Let $\mathcal{A} = (S', A')$ be a reaction system and let W_0 be initial state of $\mathcal{A}$ where*

- *$S' = \{[x, i], [x, i, k]' \mid x \in S, 1 \leq i, k \leq n\}$, and*
- *A' consists of the following reactions.*
  - *For any pc-reaction $\rho : (R_\rho, I_\rho, \Pi_\rho)$ of component A_i, there is reaction $\rho' : (R_{\rho'}, I_{\rho'}, P_{\rho'})$ in A' with $R_{\rho'} = \{[x, i] \mid x \in R_\rho\}$, $I_{\rho'} = \{[y, i] \mid y \in I_\rho\}$, $P_{\rho'} = \{[x, i, k]' \mid (x, k) \in \Pi_\rho, 1 \leq k \leq n\}$.*

• *For every* $x \in S$ *and* $1 \leq i, k \leq n$ *there is a reaction* $\rho_{[x,i,k]'} : (\{[x,i,k]'\}, \{[x,k]\}, \{[x,k]\})$ *in* A'.

- W_0 *consists of all reactants* $[x,h]$ *where* $x \in S$ *and* $[x,h]$ *is an element of* $D_{h,0}$, $1 \leq h \leq n$.

Then for any j, $j \geq 0$, *under transition* $tr : \bar{D}_j \Longrightarrow \bar{D}_{j+1}$ *in the state sequence* $\bar{D}_0, \bar{D}_1, \ldots, \bar{D}_j, \bar{D}_{j+1}, \ldots$ *of* Δ *there is an active communication link from component* A_i *to component* A_k *of* Δ *if and only if for some* $x \in S$ *there is a reactant* $[x,i,k]' \in S'$ *which is a product of an enabled reaction of* $\mathcal{A}$ *on* W_{2j} *in transition* $W_{2j} \Longrightarrow W_{2j+1}$ *of the state sequence* $W_0, W_1, \ldots$ *of* $\mathcal{A}$.

This statement can be proven by modifying the proof of Theorem 1, we leave the details to the reader.

3.2 Communication by Reactions

Under operation, the architecture of the cdcR(p) system remains unchanged in the sense that the set of reactions of the component does not change. An interesting question is the following: What can we say about communicating reaction systems where the current sets of reactions of the components are allowed to change from state to state. One possible variant of this model is where the (successfully) performed reactions can be communicated to the other components and if a reaction is available at some component in some state then it had to be performed at some component in the previous state (except the case of the initial state). This type of cdcR systems can be considered as a dynamically evolving system and represents a communication model where rules and not data are communicated.

Definition 10. *A cdcR system communicating by reactions (a cdcR(r) system, for short) of degree* n, $n \geq 1$, *is a triplet* $\Delta = (n, S, \mathcal{R})$ *where*

- n *is the number of components,*
- S *is a finite nonempty set, called the background set of* Δ,
- $\mathcal{R}$ *is a finite nonempty set of extended reactions of type rc (rc-reactions, for short), where*
  • *each rc-reaction is of the form* $\rho : (R_\rho, I_\rho, P_\rho); target(\rho)$,
  • R_ρ, I_ρ, P_ρ *are nonempty subsets of* S, *the set of reactants, the set of inhibitors, and the set of products of the rc-reaction, respectively,*
  • $target(\rho) \subseteq \{1, \ldots, n\}$ *is a nonempty set, the set of indices (labels) of the target components to which the rc-reaction is communicated.*

The components are labeled by numbers i, $1 \leq i \leq n$.

For an rc-reaction $\rho = (R_\rho, I_\rho, P_\rho); target(\rho)$, triplet (R_ρ, I_ρ, P_ρ) is called its core and is denoted by $core(\rho)$. For a nonempty set $\mathcal{R}' \subseteq \mathcal{R}$ we define $core(\mathcal{R}') = \{core(\rho) \mid \rho \in \mathcal{R}'\}$. An rc-reaction ρ is enabled for a nonempty subset U of S if $core(\rho)$ is enabled for U; the result of performing ρ on U means the result of performing $core(\rho)$ on U. Notations $en_\rho(U)$, $res_\rho(U)$, and $en_{\mathcal{R}'}(U)$, $res_{\mathcal{R}'}(U)$

where ρ is an rc-reaction and R' is a set of rc-reaction systems are used in the usual manner.

If no confusion arises, from now on ρ will be called the label of reaction ρ as well.

Next we define the operation of cdcR(r) systems. These systems work with changing their configurations, i.e. changing the current reaction sets and the current sets of reactants that are at the disposal of the components. While the behavior of cdcR(p) systems can be represented by the state sequences, in case of cdcR(r) systems we speak of configuration sequences, since reaction sets are allowed to be changed as well.

Definition 11. *Let $\Delta = (n, S, \mathcal{R})$, $n \geq 1$, be a cdcR(r) system with n components. Let $\bar{C}_0$ be the initial configuration of Δ where $\bar{C}_0 = ((A_{1,0}, D_{1,0}) \ldots, (A_{n,0}, D_{n,0}))$ with $A_{i,0} \subseteq \mathcal{R}$ (the initial rc-reaction set of component i) and $D_{i,0} \subseteq S$ (the initial reactant set of component i), $1 \leq i \leq n$. The pair $(A_{i,0}, D_{i,0})$ is called the initial configuration of component i.*

The configuration sequence $\bar{C}_0, \bar{C}_1, \ldots$ of Δ, where $\bar{C}_j = ((A_{1,j}, D_{1,j}) \ldots, (A_{n,j}, D_{n,j}))$, $j \geq 0$, is defined as follows:

For each component i, $1 \leq i \leq n$, for each j, $j \geq 0$ and for every subsequent configurations $(A_{i,j}, D_{i,j})$, $(A_{i,j+1}, D_{i,j+1})$ of component i the following hold:

- $A_{i,j+1} = \{\rho \in \mathcal{R} \mid i \in target(\rho), \rho \in A_{k,j}, en_{core(\rho)}(D_{k,j}), 1 \leq k \leq n\}$ *and*
- $D_{i,j+1} = res_{core(A_{i,j})}(D_{i,j})$

That is, after performing the reactions that are enabled for the current reactant sets at the components, the products stay with the components and those reactions that were enabled for the reactant set are communicated. This means that these reactions are added to the reaction sets of their target components. (Notice that the sender component can be a target component as well). The new set of reactions of the component consists of all reactions that were obtained by communication. (Thus, those reactions that were not enabled for the reactant set are erased from the set of reactions of the component.)

We give an example for a cdcR(r) system.

Example 2. Let $\Delta = (3, S, \mathcal{R})$ be a cdcR(r) system where $S = \{a, b, c, d\}$ and $\mathcal{R}$ is defined as follows. Let

$$\begin{aligned}
\mathcal{R} = \{&\rho_1 : (\{a, b\}, \{d\}, \{a\}); \{1, 2\},\\
&\rho_2 : (\{b\}, \{d\}, \{b\}); \{1, 2\},\\
&\rho_3 : (\{a, b\}, \{c\}, \{c\}); \{2, 3\},\\
&\rho_4 : (\{a\}, \{c\}, \{a\}); \{2, 3\},\\
&\rho_5 : (\{a, c\}, \{b\}, \{a\}); \{3, 1\},\\
&\rho_6 : (\{a\}, \{d\}, \{b\}); \{3, 1\}\}.
\end{aligned}$$

Let the initial configuration of Δ, $\bar{C}_0 = ((A_{1,0}, D_{1,0}), (A_{2,0}, D_{2,0}), (A_{3,0}, D_{3,0}))$ be given as follows. Let $A_{1,0} = \{\rho_1, \rho_2\}$, $A_{2,0} = \{\rho_3, \rho_4\}$, and $A_{3,0} = \{\rho_5, \rho_6\}$. Let $D_{1,0} = \{a, b\}$, $D_{2,0} = \{a, b\}$, $D_{3,0} = \{a, c\}$, i.e. the same initial states and sets of reactants as in Example 1.

The new configuration $\bar{C}_1$ of Δ will be the following. It can easily be seen that each reaction can be performed at each component, thus the new rc-reaction sets will be the following. The first component will have rc-reactions $\rho_1, \rho_2, \rho_5, \rho_6$, the second component will have rc-reactions $\rho_3, \rho_4, \rho_1, \rho_2$, and the third component will have rc-reactions $\rho_5, \rho_6, \rho_3, \rho_4$. The new states will be $\{a, b\}$, $\{a, c\}$, $\{a, b\}$, respectively. Repeating the procedure, the state of the first component will be $\{a, b\}$, the state of the second component will be $\{a, b\}$, and the third component will have state $\{a, b\}$ as well.

As with cdcR(p) systems, to every cdcR(r) system Δ we can construct an R system $\mathcal{A}$ which represents Δ.

Theorem 3. *Let $\Delta = (n, S, \mathcal{R})$, $n \geq 1$ be a cdcR(r) system of degree n, and let $Lab_R = \{l_\rho \mid \rho \in \mathcal{R}\}$ be the set of labels associated to the elements of $\mathcal{R}$; Lab_R and S are disjoint sets.*

Let $\sigma = \bar{C}_0, \bar{C}_1, \ldots$ be the configuration sequence of Δ starting from initial configuration $\bar{C}_0$, where $\bar{C}_j = ((A_{1,j}, D_{1,j}) \ldots, (A_{n,j}, D_{n,j}))$, $j \geq 0$.

We can construct a reaction system $\mathcal{A} = (S', A')$, give initial state W_0 of $\mathcal{A}$ and mappings h_i, g_i, $1 \leq i \leq n$ such that for every pair $(A_{i,j}, D_{i,j})$, $j \geq 0$, in the configuration sequence σ it holds that $h_i(W_j) = D_{i,j}$ and $g_i(W_j) = Lab_{A_{i,j}}$ where $Lab_{A_{i,j}}$ denotes the set of labels of rc-reactions that are elements of $A_{i,j}$ and $W_0, W_1, \ldots$ is the state sequence of $\mathcal{A}$ starting from W_0.

Proof. Let us define $\mathcal{A} = (S', A')$ as follows. Let $S' = \{[a, i] \mid a \in (S \cup Lab_R), 1 \leq i \leq n\}$. To each rc-reaction $\rho : (R_\rho, I_\rho, P_\rho); target(\rho)$ in $\mathcal{R}$ and for each i, $1 \leq i \leq n$, we define a reaction $(\rho', i) : (\{[l_\rho, i]\} \cup \{[a, i] \mid a \in R_\rho\}, \{[b, i] \mid b \in I_\rho\}, \{[c, i] \mid\in P_\rho\} \cup \{[l_\rho, k] \mid k \in target(\rho)\})$.

Let $W_0 = \bigcup_{i=1}^{n}(\{[l_\rho, i] \mid l_\rho \in Lab_R, \rho : (R_\rho, I_\rho, P_\rho); target(\rho) \in A_{i,0}\} \cup \{[b, i] \mid b \in D_{i,0}\})$.

Let us define mapping $h_i : 2^{S'} \to 2^S$, $1 \leq i \leq n$ as follows. For $U \subseteq S'$ with $U \cap S' \neq \emptyset$, let $h_i(U) = \{x \mid [x, i] \in U\}$, otherwise let $h_i(U) = \emptyset$. (Notice that if $U = \{[x, i], [y, j]\}$ where $j \neq i$, then $h_i(U) = \{x\}$.) Let mapping $g_i : 2^{S'} \to 2^{Lab_R}$, $1 \leq i \leq n$ be defined as follows. For $V \subseteq S'$ and $V \cap S' \neq \emptyset$ let $g_i(V) = \{l_\rho \mid [l_\rho, i] \in V, V \subseteq Lab_R\}$, otherwise let $g_i(V) = \emptyset$.

By definition, it is obvious that $h_i(W_0) = D_{i,0}$ and $g_i(W_0) = Lab_{A_{i,0}}$, where $Lab_{A_{i,0}}$ denotes the labels of reactions in $A_{i,0}$.

Suppose now that for any fixed i, and up to certain j, $j \geq 1$ for $(A_{i,j}, D_{i,j})$ in the configuration sequence of Δ it holds that $h_i(W_j) = D_{i,j}$ and $g_i(W_j) = Lab_{A_{i,j}}$ where $Lab_{A_{i,j}}$ denotes the set of labels of reactions that are elements of $A_{i,j}$, and W_j is the jth element in the state sequence of δ starting from its initial state W_0. We show now that the statement holds for $j + 1$ as well.

Notice that due to the form of the reactions of $\mathcal{A}$, for any j, where $j \geq 1$, $[l_\rho, i]$ appears in W_j if and only if it was obtained as a product in the previous step by some reaction of $\mathcal{A}$. By the reactions of $\mathcal{A}$ this is possible if and only if (R_ρ, I_ρ, P_ρ) was performed at some component of Δ and ρ was communicated to component i. Thus, reactants of the form $[l_\rho, i]$ in W_j and reactions in $A_{i,j}$ correspond to each other. Analogously, any reactant of the form $[b, i]$ occurs in

W_j if and only if it is an element of $D_{i,j}$. Applying reactions of $\mathcal{A}$ to W_j, elements of W_{j+1} will be of the form $[\gamma, i]$ and $[c, i]$ where $\gamma \in Lab_{A_{i,j+1}}$ and $c \in S$ meet the previously listed conditions. Notice that labels of reactions of Δ are reactants of $\mathcal{A}$ that indicate the simulation of a reaction in Δ with a reaction of $\mathcal{A}$. Thus, that the statement of the theorem holds.

Analogously to Theorem 1, the previous statement has a direct consequence. So the proof is left to the reader.

Corollary 2. *Let Δ be a cdcR(r) system of degree n, $n \geq 1$, and let $\mathcal{A}$ be an R system given as in Theorem 3. Let $\bar{C}_0$ be the initial configuration of Δ and let W_0 be the initial state of $\mathcal{A}$ given as in Theorem 3. Then rc-reaction ρ occurs at component i in the mth element of state sequence of Δ starting from $\bar{C}_0$ if and only if reactant $[l_\rho, i]$ occurs in the mth element of state sequence of $\mathcal{A}$ starting from W_0, where $m \geq 1$.*

As for cdcR(p) systems, the reaction system $\mathcal{A}$ constructed to cdcR(r) system Δ in Theorem 3 can be called the flattened reaction system of Δ and we can formulate an occurrence problem to cdcR(r) systems as follows. For a given cdcR(r) system $\Delta = (n, S, \mathcal{R})$, $n \geq 1$, the problem whether an rc-reaction $\rho \in \mathcal{R}$ occurs at the ith component at the mth element of the state sequence of Δ starting with some initial configuration $\bar{C}_0$ is called the occurrence problem of cdcR(r) systems. By Theorem 3 and Corollary 2, and by [10,12] we may state that the occurrence problem of cdcR(r) systems for some fixed values of m is NP-complete and it is a PSPACE-problem when m is given as input.

Analogously to cdcR(p) systems, we define the flattened reaction system of cdcR(r) systems Δ.

Definition 12. *Let $\Delta = (n, S, \mathcal{R})$, $n \geq 1$ be a cdcR(r) system of degree n, and let $Lab_R = \{l_\rho \mid \rho \in \mathcal{R}\}$ be a set of labels associated to the elements of $\mathcal{R}$. Let Lab_R and S be disjoint sets. Let us define reaction system $\mathcal{A} = (S', A')$ as follows. Let $S' = \{[a, i] \mid a \in (S \cup Lab_R), 1 \leq i \leq n\}$. To each rc-reaction $\rho : (R_\rho, I_\rho, P_\rho); target(\rho)$ in $\mathcal{R}$ and for each i, $1 \leq i \leq n$, we define a reaction $(\rho', i) : (\{[l_\rho, i]\} \cup \{[a, i] \mid a \in R_\rho\}, \{[b, i] \mid b \in I_\rho\}, \{[c, i] \mid \in P_\rho\} \cup \{[l_\rho, k] \mid k \in target(\rho)\})$. $\mathcal{A}$ has no more reactions. Then $\mathcal{A}$ is called the flattened reaction system of cdcR(r) system Δ.*

We have shown that both cdcR(p) systems and cdcR(r) systems can be flattened, i.e. we can construct simulating reaction systems to both types of cdcR systems. To obtain the simulating reaction system, either we indicated the location of the reactant or we indicated both the location of the reactant and the location of the reaction in the set of new reactants. In the case of cdcR(r) systems, we added the labels of rc-reactions to the reactant set of the reactions. Studying the proofs, the reader may notice that the simulating reaction systems are similar. Based on this observation, we show that to any cdcR(r) system we can construct a cdcR(p) system such that there exists a reaction system which is the flattened version of both.

Theorem 4. *Let $\Delta = (n, S, \mathcal{R})$, $n \geq 1$ be a cdcR(r) system of degree n and let $\mathcal{A}$ be the flattened reaction system of Δ given as in Definition 12. Then there exists a cdcR(p) system Δ' such that for its flattened reaction system $\mathcal{A}'$, given as in Definition 7, $\mathcal{A} = \mathcal{A}'$ holds.*

Proof. Let us consider $\Delta = (n, S, \mathcal{R})$, $n \geq 1$ and let $Lab_R = \{l_\rho \mid \rho \in \mathcal{R}\}$ be a set of labels associated to the elements of $\mathcal{R}$. Let Lab_R and S be disjoint sets. By Definition 12 the flattened reaction system $\mathcal{A}$ of Δ is defined as follows: $\mathcal{A} = (S', A')$ where $S' = \{[a,i] \mid a \in (S \cup Lab_R), 1 \leq i \leq n\}$. To each rc-reaction $\rho : (R_\rho, I_\rho, P_\rho); target(\rho)$ in $\mathcal{R}$ and for each i, $1 \leq i \leq n$, there is a reaction $(\rho', i) : (\{[l_\rho, i]\} \cup \{[a,i] \mid a \in R_\rho\}, \{[b,i] \mid b \in I_\rho\}, \{[c,i] \mid \in P_\rho\} \cup \{[l_\rho, k] \mid k \in target(\rho)\})$. $\mathcal{A}$ has no more reactions.

Let us define cdcR(p) system Δ' as follows. Let $\Delta' = (S', A'_1, \ldots, A'_n)$, $n \geq 1$, where $S' = \{[a,i] \mid a \in S, 1 \leq i \leq n\} \cup \{[l_\rho, i] \mid \rho \in \mathcal{R}, 1 \leq i \leq n\}$. Let A'_i be defined as follows: for $\rho : (R_\rho, I_\rho, P_\rho); target(\rho)$ in $\mathcal{R}$ we define pc-reaction $\rho' : (\{l_\rho\} \cup R_\rho, I_\rho, \{[c,i] \mid c \in P_\rho\} \cup \{l_\rho(j) \mid j \in target(\rho)\})$.

It is easy to see that after performing the pc-reaction ρ', elements of S that are products in ρ stay with the component, while the label of ρ, l_ρ, is communicated to those components that are given as targets of ρ in Δ.

Now let us construct the flattened version of Δ', given in Definition 7, denoted by $\mathcal{A}'$. Then for each reaction ρ' of Δ', see above, we obtain reaction (ρ'', i) : $(\{[l_\rho, i]\} \cup \{[a,i] \mid a \in R_\rho\}, \{[b,i] \mid b \in I_\rho\}, \{[c,i] \mid c \in P_\rho\} \cup \{[l_\rho, k] \mid k \in target(\rho)\})$. Then it is easy to see that $\mathcal{A}' = \mathcal{A}$ holds.

4 Conclusions

In this paper we introduced new variants of networks of reaction systems where the components communicate with each other by sending products or reactions. We proved that these networks can be represented by single reaction systems (flattened reaction systems), and discussed some aspects of communication in these networks. We pointed out a connection between the occurrence of a reactant (a reaction) at some component of the cdcR(p) system (cdcR(r) system) at some step of the operation and the occurrence of the corresponding reactant in the same step of the operation of the corresponding flattened reaction system. Occurrence problems and their complexity for reaction systems have been studied in [10,12] and were shown to be NP-complete (or PSPACE-complete) problems, depending on how the problem is formulated. These studies and results can be interpreted in terms of cdcR(p) systems (cdcR(r)) systems. In the future, we plan to study the connections between R systems and P systems (see, for example [1,2]). Further types of direct communication protocols, dynamic behavior would also be of interest to investigate.

Acknowledgment. The authors thank the reviewers for their valuable comments. The work of Erzsébet Csuhaj-Varjú was supported by the National Research, Development, and Innovation Office - NKFIH, Hungary, Grant no. K 120558. The work of Pramod Kumar Sethy was supported by project "Integrált kutatói utánpótlás-képzési program az informatika és számítástudomány diszciplináris területein", EFOP 3.6.3-VEKOP-16-2017-00002, a project supported by the European Union and co-funded by the European Social Fund.

References

1. Alhazov, A.: P systems without multiplicities of symbol-objects. Inf. Process. Lett. **100**(3), 124–129 (2006)
2. Alhazov, A., Aman, B., Freund, R., Ivanov, S.: Simulating R Systems by P Systems. In: Leporati, A., Rozenberg, G., Salomaa, A., Zandron, C. (eds.) CMC 2016. LNCS, vol. 10105, pp. 51–66. Springer, Cham (2017). https://doi.org/10.1007/978-3-319-54072-6_4
3. Bottoni, P., Labella, A., Rozenberg, G.: Reaction systems with influence on environment. J. Membr. Comput. **1**(1), 3–19 (2019). https://doi.org/10.1007/s41965-018-00005-8
4. Bottoni, P., Labella, A., Rozenberg, G.: Networks of reaction systems. Int. J. Found. Comput. Sci. **31**(1), 53–71 (2020)
5. Castellanos, J., Martín-Vide, C., Mitrana, V., Sempere, J.M.: Solving NP-complete problems with networks of evolutionary processors. In: Mira, J., Prieto, A. (eds.) IWANN 2001. LNCS, vol. 2084, pp. 621–628. Springer, Heidelberg (2001). https://doi.org/10.1007/3-540-45720-8_74
6. Csuhaj-Varjú, E., Kelemen, J., Păun, G.: Grammar systems with wave-like communication. Comput. Artif. Intell. **15**(5), 419–436 (1996)
7. Csuhaj-Varjú, E., Salomaa, A.: Networks of parallel language processors. In: Păun, G., Salomaa, A. (eds.) New Trends in Formal Languages. LNCS, vol. 1218, pp. 299–318. Springer, Heidelberg (1997). https://doi.org/10.1007/3-540-62844-4_22
8. Ehrenfeucht, A., Rozenberg, G.: Basic notions of reaction systems. In: Calude, C.S., Calude, E., Dinneen, M.J. (eds.) DLT 2004. LNCS, vol. 3340, pp. 27–29. Springer, Heidelberg (2004). https://doi.org/10.1007/978-3-540-30550-7_3
9. Ehrenfeucht, A., Rozenberg, G.: Reaction systems. Fundam. Informaticae **75**(1–4), 263–280 (2007)
10. Formenti, E., Manzoni, L., Porreca, A.E.: On the complexity of occurrence and convergence problems in reaction systems. Nat. Comput. **14**(1), 185–191 (2014). https://doi.org/10.1007/s11047-014-9456-3
11. Hopcroft, J.E., Motwani, R., Ullman, J.D.: Introduction to Automata Theory, Languages, and Computation, 3rd edn. Addison-Wesley, Pearson International Edition (2007)
12. Salomaa, A.: Functional constructions between reaction systems and propositional logic. Int. J. Found. Comput. Sci. **24**(1), 147–160 (2013)

Generalized Forbidding Matrix Grammars and Their Membrane Computing Perspective

Henning Fernau[1(✉)], Lakshmanan Kuppusamy[2], and Indhumathi Raman[3]

[1] Fachbereich 4 – Abteilung Informatikwissenschaften, Universität Trier, 54286 Trier, Germany
fernau@uni-trier.de

[2] School of Computer Science and Engineering, VIT University, Vellore 632 014, India
klakshma@vit.ac.in

[3] Department of Applied Mathematics and Computational Sciences, PSG College of Technology, Coimbatore 641 004, India
ind.amcs@psgtech.ac.in

Abstract. Matrix grammars are one of the first approaches ever proposed in regulated rewriting, prescribing that rules have to be applied in a certain order. In traditional regulated rewriting, the most interesting case shows up when all rules are context-free. Typical descriptional complexity measures incorporate the number of nonterminals or the length, i.e., the number of rules per matrix. When viewing matrices as program fragments, it becomes natural to consider additional applicability conditions for such matrices. Here, we focus on forbidding sets, i.e., a matrix is applicable to a sentential form w only if none of the words in its forbidding set occurs as a subword in w. This gives rise to further natural descriptional complexity measures: How long could words in forbidding sets be? How many words could be in any forbidding set? How many matrices contain non-empty forbidding contexts? As context-free grammars with forbidding sets are known as generalized forbidding grammars, we call this variant of matrix grammars also generalized forbidding. In this paper, we attempt to answer the above four questions while studying the computational completeness of generalized forbidding matrix grammars. We also link our research to processing strings with membrane computing and discuss appropriate variations of P systems.

Keywords: Generalized forbidding grammars · Matrix grammars · Computational completeness · Descriptional complexity · P systems

1 Introduction

Matrix Grammars. If rules of a context-free grammar are grouped into finite sequences called matrices, we arrive at *matrix grammars* (originally introduced

R. Freund et al. (Eds.): CMC 2020, LNCS 12687, pp. 31–45, 2021.
https://doi.org/10.1007/978-3-030-77102-7_3

by S. Ábrahám [1] on linguistic grounds); when a matrix is chosen to be applied to a sentential form, all rules in the sequence are applied in the given order. Matrix grammars with appearance checking, having three nonterminals, are computationally complete, i.e., they characterize RE [6]. However, the lengths of the matrices are unbounded. It is not clear how to restrict the length while still bounding the number of nonterminals. Note that matrix grammars without appearance checking (abbreviated as MAT) are not computationally complete [14].

Semi-conditional Grammars. In 1985, Gh. Păun introduced another variant of regulated rewriting, so-called semi-conditional grammars [22]. To each (context-free) rule, a permitting and a forbidden string are associated that govern the applicability of said rule. In [22], also combinations with other regulation mechanisms were investigated. For our studies, in particular the variation leading to the language classes $\mathcal{KM}^{\lambda}(i,j)$ is of special interest where to each matrix (containing sequences of possibly erasing context-free rules), a permitting string w_1 of length at most i and a forbidden string w_2 of length at most j are associated. Such a matrix is only applicable to the sentential form w if w_1 is a substring of w and if w_2 does not occur as a substring of w. We have integrated the result $\mathcal{KM}^{\lambda}(0,2) = \mathsf{RE}$ of Theorem 4.4 in [22] into our survey Table 1.

Generalized Forbidding Grammars. In 1990, A. Meduna [19] introduced the following modification of semi-conditional grammars: disallowing permitting strings but allowing a number of forbidden ones. If in a context-free grammar, each rule is associated with a set F of strings (called the forbidding set), then such a grammar is called a *generalized forbidding* (GF) grammar. A rule can be applied to a sentential form w if none of the strings in the associated forbidding sets occur as a substring in w. There are four main parameters that describe the size of a GF grammar, namely, (1) d, the maximum length of strings in the forbidding sets, (2) i, the maximum cardinality of the forbidding sets, (3) n, the number of nonterminals used in the grammar, and (4) c, the number of rules with nonempty forbidding set. The family of languages described by a GF grammar of size (at most) (d,i,n,c) is denoted by $\mathsf{GF}(d,i,n,c)$.

Generalized Forbidding Matrix Grammars. In this paper, we combine the studies of $\mathcal{KM}$- and GF-grammars by introducing *generalized forbidding matrix* (GFM) grammars. Here, forbidding sets are associated to matrices of context-free rules. Apart from the parameters d,i,n discussed for GF-grammars, we consider the number m of conditional matrices (ignoring matrices that have empty forbidding sets) and an upper bound l on the length of the matrices. This leads to language families like $\mathsf{GFM}(d,i,n;m,l)$. Notice that $\mathsf{GFM}(d,i,n;m,1) = \mathsf{GF}(d,i,n,m)$.

Survey of Main Results. We review known and new descriptional complexity results on computational complete $\mathsf{GFM}(d,i,n;m,l)$ in Table 1. A star indicates that the corresponding construction does not allow for any bounds on this parameter. We have grouped our results according to general constructions

Table 1. Computational completeness results for GFM grammars

Degree	Index	# Nontermin	# Matrices	Matrix length	Reference
d	i	n	m	l	
1	2	*	*	1	[22]
*	2	3	*	*	Thm. 1
1	1	*	*	2	Thm. 2
2	6	8	6	1	[7,8]
2	5	9	7	1	[8]
2	4	7	8	1	[8]
2	3	20	18	1	[7,8]
2	3	7	2	3	Thm. 6
2	5	6	3	2	Thm. 7
2	5	6	2	3	Thm. 8
2	3	7	4	2	Thm. 9
2	3	8	3	2	Thm. 10
2	2	8	6	2	Thm. 11

(where some parameters are not bounded), followed by descriptional complexity results on GF grammars,[1] interpreted as results on GFM grammars as described above, and finally we obtain a number of (new) descriptional complexity results on GFM grammars. Let us mention a couple of special features here:

- With GF grammars (i.e., GFM grammars with $l = 1$), we are not aware of any computational completeness results with six or less nonterminals. With matrix length two, we do get this type of results for GFM grammars.
- The number of conditional matrices is significantly higher with matrix length one compared to allowing longer matrices.
- At first glance, it might look unfair to compare the number of conditional rules in GF grammars with the number of conditional matrices in GFM grammars; however, observe that the forbidding sets are attached to matrices within GFM grammars which means, in a sense, that they are associated to the first rule in the matrices and are not used for the other rules of the matrices. In this interpretation, the number of conditional matrices equals the number of conditional rules.

Connections to Membrane Computing. Some of our results concern matrix length two. In particular, this concerns Theorems 2, 7, 9, 10, and 11. Such systems

[1] Notice that computational completeness results have quite some history in the literature of GF grammars; we only refer to [7,17–20] in some historical order. In the long version of [7] which is appearing in *Discrete Applied Mathematics* [8], we obtained some further improved results of the conference version [7].

could be easily interpreted as a special form of membrane computing, an area of research started by Gh. Păun two decades ago with the ground-breaking article [23]. The resulting membrane computing systems have quite a particular structure: there is only one outer membrane that contains several inner membranes. No inner membrane contains (recursively) further inner membranes, i.e., the underlying tree structure has a very limited depth. For each matrix of length two, there is an inner membrane. The first rules of all matrices are processed in the outer membrane. In particular, this means that all matrices of length one (typically, these are corresponding to the unconditional context-free rules) are worked in the outer membrane, where also the processing starts with a single start symbol. Most of our simulations start out with some form of Geffert normal form. This means that all rules that correspond to the phase one of the Geffert normal form grammar are simulated within the outer membrane alone. In the simulation of the second phase, the inner membranes will come into play. The forbidding sets are then kind of filters attached to rules that could send sentential forms to an inner membrane. This correlates to the universality results obtained in P systems with forbidding contexts having a membrane structure of depth two [11]. Yet, the numbers of nonterminals or production rules are unbounded there [2,11]. Also in P systems with promoters and inhibitors (see [3,15]), the reaction of chemicals (a multiset of objects) can happen in the presence and absence of certain chemicals with or without catalysts (i.e., enzymatic proteins).

2 Generalized Forbidding Matrix Grammars

Let us now formally introduce the main subject of our study. A *generalized forbidding matrix (GFM) grammar* is a quadruple $G = (V, T, M, S)$; V is the total alphabet, $T \subset V$ is the terminal alphabet, $S \in V \setminus T$ is the start symbol and M is a set of matrices of the form $m = [(A_1 \to x_1), \ldots, (A_\ell \to x_{l_m}), F_m]$, where $A_i \in V \setminus T$, $x_i \in V^*$, $F_m \subseteq V^+$; $|\bigcup_{m \in M} F_m|, |M| < \infty$. F_m is the *forbidding set* of m and l_m is its *length.* If $F_m = \emptyset$, we may omit F and call such a matrix *unconditional*, and if in addition $l_m = 1$, we identify the matrix $[(A \to x)]$ with the context-free rule $A \to x$.

Next, we define the semantics of the GFM grammar $G = (V, T, M, S)$, i.e., the language $L(G)$ that G generates. Let $m = [(A_1 \to x_1), \ldots, (A_\ell \to x_{l_m}), F_m] \in M$. If $x, y \in V^*$, then $x \Rightarrow_m y$ holds if

- no string from F_m is a substring of x;
- there exist $m + 1$ strings $z_i \in V^*$, with $i = 0, \ldots, m$, such that
  - $z_0 = x$, $z_m = y$;
  - for $i = 1, \ldots, m$: z_i is obtained from z_{i-1} by applying the context-free rule $A_i \to x_i$, which means that there are strings ζ_i, η_i such that $z_{i-1} = \zeta_i A_i \eta_i$ and $z_i = \zeta_i x_i \eta_i$.

From all relations $\Rightarrow_m \subseteq V^* \times V^*$, we build the *one-step derivation relation* $\Rightarrow_G$ of G as $\Rightarrow_G := \bigcup_{m \in M} \Rightarrow_m$. If $\Rightarrow_G^*$ denotes the reflexive transitive closure, we can now define $L(G) := \{w \in T^* \mid S \Rightarrow_G^* w\}$.

We are interested in the following descriptional complexity parameters of a GFM grammar $G = (V, T, M, S)$:

1. $d(G) = \max_{m \in M} \max_{w \in F_m} |w|$, i.e., the maximum length of forbidden strings; this parameter is also known as the *degree* of G.
2. $i(G) = \max_{m \in M} |F_m|$, i.e., the maximum cardinality of a forbidding set; this parameter is also known as the *index* of G.
3. $n(G) = |V \setminus T|$, i.e., the number of nonterminals;
4. $m(G) = |\{m \in M \mid F_m \neq \emptyset\}|$, i.e., the number of *conditional* matrices;
5. $l(G) = \max_{m \in M} l_m$, i.e., the maximum *length* of a matrix.

This allows us to define the language families that we are studying in this paper under aspects of computational completeness:

$$\mathsf{GFM}(d, i, n; m, l) := \{L(G) \mid d(G) \leq d, i(G) \leq i, n(G) \leq n, m(G) \leq m, l(G) \leq l\}.$$

Of course, we are considering GFM grammars only in this definition. Notice that we put a semi-colon after the first three parameters in order to separate the parameters that are identical with forbidding context-free grammars [7] and those that are specific to matrix grammars.

3 Results for Unbounded Context Lengths

We start with some results from [6] where it has been shown that any recursively enumerable language can be generated by some graph-controlled grammar with only two nonterminals, improving on earlier works as [5,12,21] (in some chronological order). We can now adapt the construction of Corollary 6 in [6] to obtain:

Theorem 1. $RE = GFM(*, 2, 3; *, *)$.

In other words, only the number of forbidden strings per set and the number of nonterminals is bounded. With an alternative interpretation, we can also get a computational completeness result where the number of nonterminals is unbounded, while the maximum length of a matrix is now at most two.

Theorem 2. $RE = GFM(1, 1, *; *, 2)$.

We think that both results are interesting, because they give a sort of lower bound on what can be achieved if we delimit all the parameters.

Interestingly enough, the second theorem closes a gap in the theory of $\mathcal{KM}$-grammars, cf. [22, Theorem 4.4], as we can conclude:

Corollary 1. $CF = \mathcal{KM}^{\lambda}(0, 0) \subsetneq \mathcal{KM}^{\lambda}(0, 1) = RE$.

Our proofs follow rather traditional ways how to simulate graph-controlled (or programmed) grammars by matrix grammars, as can be also found in the textbook [4]. Therefore, we omit further proof details here. The main difference between the proofs of both theorems is that for the first proof, we use (long) sequences of the same symbol, say, C in order to encode the current state (graph node), while for the second proof, we use special symbols. Hence, the first proof allows to delimit the number of nonterminals to three, at the expense of matrices of unbounded length, while the second one allows to delimit the length of the matrices to two, but there is no bound anymore on the number of nonterminals.

4 Normal Forms of Phrase Structure Grammars

All of our more refined constructions start with a phrase structure grammar in a certain normal form. In fact, all of these normal forms originate from the seminal work of V. Geffert [13]. The best known of these normal forms, often referred to as *Geffert normal form* (GNF), uses five non-terminals S, A, B, C, D and two erasing non-context-free rules $AB \to \lambda$ and $CD \to \lambda$. We will call this normal form as (5, 2)-GNF, highlighting the number of nonterminals and the number of non-context-free rules in this way, because they are characteristic for the normal forms that Geffert found. The derivation of a type-0 grammar in (5, 2)-GNF proceeds in two phases, where the first phase splits into two stages. In phase one, stage one, rules of the form $S \to uSa$ are used, with $u \in \{A, C\}^+$, $a \in T$. In stage two of phase one, rules of the form $S \to uSv$ are used, with $u \in \{A, C\}^+$ and $v \in \{B, D\}^*$.[2] Also, rules of the form $S \to uv$ are available (see [13]) that prepare the transition into phase two, where the two erasing non-context-free rules $AB \to \lambda$ and $CD \to \lambda$ are used exclusively until a terminal string is derived.

Accordingly, a type-0 grammar is said to be in (4, 1)-GNF if it has exactly four nonterminals S, A, B, C and a single non-context-free erasing rule of the form $ABC \to \lambda$. Geffert has shown in [13] that this normal form is obtained from (5, 2)-GNF by applying the morphism $A \mapsto AB$, $B \mapsto C$, $C \mapsto A$ and $D \mapsto BC$ to all context-free rules.

T. Masopust and A. Meduna [17] came up with the following modification of (5, 2)-GNF that we suggested calling MMNF in [7]: Let $\tilde{G} = (\tilde{N}, T, \tilde{P}, \tilde{S})$ be a grammar in (5, 2)-GNF with $\tilde{N} = \{\tilde{S}, A, B, C, D\}$. Then, there is a grammar $G = (\{S, 0, 1, \$\}, T, P_u \cup \{0\$0 \to \$, 1\$1 \to \$, \$ \to \lambda\}, S)$, with P_u containing only unconditional rules of the form

- $S \to h(u)Sa$ if $S \to uSa \in \tilde{P}$,
- $S \to h(u)Sh(v)$ if $S \to uSv \in \tilde{P}$,
- $S \to h(u)\$h(v)$ if $S \to uv \in \tilde{P}$,

where $h : \{A, B, C, D\}^* \to \{0, 1\}^*$ is a homomorphism defined by $h(A) = h(B) = 00$, $h(C) = 01$, and $h(D) = 10$, such that $L(G) = L(\tilde{G})$.

In our present applications, it is useful to use yet another normal form, which we call *modified Masopust-Meduna normal form*, or MMMNF for short. This normal form starts out again with (5, 2)-GNF but uses (instead of h) the coding h' with $h'(A) = 00$, $h'(B) = 11$, $h'(C) = h'(D) = 01$, and as non-context-free erasing rules, we have $0\$1 \to \$$ and $1\$0 \to \$$. We inherit the nice properties of MMNF, including having only four nonterminals, namely, $S, 0, 1, \$$. Let us make this normal form more prominent in the following theorem.

Theorem 3. *For each language $L \in \mathsf{RE}$ with $L \subseteq T^*$, there exists a type-0 grammar $G = (\{S, 0, 1, \$\}, T, P, S)$ in MMMNF with $L(G) = L$ such that G only has the following types of rules:*

[2] For the subtle distinction between possibly allowing or disallowing the empty word for v or u, respectively, we refer to the discussions in [7].

- $S \to uSa$, *with* $u \in \{00, 01\}^+$ *and* $a \in T$;
- $S \to uSv$ *and* $S \to u\$v$, *with* $u \in \{00, 01\}^+$ *and* $v \in \{11, 01\}^*$,
- $0\$1 \to \$$, $1\$0 \to \$$, $\$ \to \lambda$.

There is one technicality with all these normal forms. We cannot rely on rules from stage one and stage two of phase one being used in order (as intended). In fact, Geffert's correctness proof for his normal forms needed to show that such mixed applications do not lead to unwanted terminal strings. When using these normal forms in proofs of computational completeness, we have to take care of this, as well. One way of achieving this goal is to make sure that no terminal symbol is ever followed by a nonterminal symbol. In particular when simulating derivations of this kind of normal form with grammars that rely on forbidding strings, this test does not work out. The easiest way to resolve this problem is to explicitly split the first two stages by using a new additional start symbol S' as discussed in the following theorem. As we somewhat artificially introduce another nonterminal, we call this variation $(4+1, 1)$-GNF.

Theorem 4. *For each language* $L \in \mathsf{RE}$ *with* $L \subseteq T^*$, *there exists a type-0 grammar* $G = (\{S', S, A, B, C\}, T, P, S')$ *in* $(4+1, 1)$*-GNF with* $L(G) = L$ *such that* G *only has the following types of rules:*

- $S' \to uS'a$, $u \in \{A, AB\}^+$, $a \in T$,
- $S' \to uSv$, $S \to uSv$, $S' \to uv$, $S \to uv$, $u \in \{A, AB\}^+$, $v \in \{BC, C\}^*$,
- $ABC \to \lambda$.

Similarly, we introduce the variation MMMNF+1 that uses an additional nonterminal S' as the new start symbol. This idea leads to the next result.

Theorem 5. *For each language* $L \in \mathsf{RE}$ *with* $L \subseteq T^*$, *there exists a type-0 grammar* $G = (\{S', S, 0, 1, \$\}, T, P, S')$ *in MMMNF+1 with* $L(G) = L$ *such that* G *only has the following types of rules:*

- $S' \to uS'a$, *with* $u \in \{00, 01\}^+$ *and* $a \in T$;
- $S' \to uSv$, $S \to uSv$, $S' \to u\$v$ *and* $S \to u\$v$, *with* $u \in \{00, 01\}^+$ *and* $v \in \{11, 01\}^*$,
- $0\$1 \to \$$, $1\$0 \to \$$, $\$ \to \lambda$.

5 When All Parameters Are Bounded

In this section, we make quite some use of the $(4+1, 1)$-GNF and the MMMNF+1 as described in Theorems 4 and 5. This allows us to prove quite a number of computational completeness results for GFM grammars that bound all the parameters.

Theorem 6. $\mathsf{GFM}(2, 3, 7; 2, 3) = \mathsf{RE}$.

$m1 = [(B \to \lambda), (A \to \$), (C \to \#), \{S, AC, \#\}]$
$m2 = [(\$ \to \lambda), (\# \to \lambda), \{\$A, \$B, \$C\}]$

Fig. 1. Simulating $ABC \to \lambda$, using rules of type $\mathsf{GFM}(2,3;7;2,3)$

Proof. We start with a type-0 grammar G in $(4+1,1)$-GNF, see Thm. 4. The GFM grammar $G' = (V, T, M, S)$ that we are going to construct inherits the terminal alphabet T from G. Moreover, $V \setminus T = \{S', S, A, B, C, \#, \$\}$ are the seven nonterminals. The context-free rules of G are simply viewed as unconditional matrices of length one and hence incorporated into G'.

Figure 1 explains how the only non-context-free rule $ABC \to \lambda$ is simulated with the help of two conditional matrices of length three. The forbidding sets of these matrices contain at most three strings of length at most two each. This proves the claimed parameter sizes. For the formal correctness of the construction, we refer to the long version of this paper. □

We are now following a different simulation idea, based on MMMNF+1.

Theorem 7. *GFM*$(2,5,6;3,2) =$ *RE.*

Proof. We start out with a type-0 grammar G in MMMNF+1. Hence, this grammar operates with the nonterminal alphabet $N = \{S', S, 0, 1, \$\}$. The context-free rules readily translate to unconditional short matrices of the simulating GFM grammar G', which has one additional nonterminal only, which is $\#$.

How the non-context-free rules of G are simulated, is explained in Fig. 2. Observe the trick that matrix $m1$ can be applied both to strings $\alpha 0\$1\beta$ and $\alpha 1\$0\beta$, leading to the same string $\alpha\#\$\#\beta$; correctness of this application (in particular, choosing the two symbols next to \$ for replacement when applying matrix $m1$) is checked with matrix $m2$.

For further details concerning the correctness of our construction, we refer to the long version of this paper. □

In the previous simulation, the last matrix only checks for the absence of several symbols, because we want to guarantee that it is applied last. If we could not guarantee this, there is the danger that matrices $m1$ and $m2$ are applied in an uncontrolled way. Alternatively, we could check the presence of \$ in these matrices, so that the deletion $\$ \to \lambda$ can happen unconditionally. This idea is implemented in the next result; the simulating matrices are shown in Fig. 3.

$m1 = [(0 \to \#), (1 \to \#), \{S, S', \#\}]$
$m2 = [(\# \to \lambda), (\# \to \lambda), \{\$0, 0\$, \$1, 1\$, \#\#\}]$
$m3 = [(\$ \to \lambda), \{0, 1, \#\}]$

Fig. 2. Simulating $0\$1 \to \$$, $1\$0 \to \$$, $\$ \to \lambda$, using rules of type $\mathsf{GFM}(2,5,6;3,2)$

$m1 = [(\$ \to \$), (0 \to \#), (1 \to \#), \{S, S', \#\}]$
$m2 = [(\$ \to \$), (\# \to \lambda), (\# \to \lambda), \{\$0, 0\$, \$1, 1\$, \#\#\}]$

Fig. 3. Simulating $0\$1 \to \$$ and $1\$0 \to \$$, using rules of type $\mathsf{GFM}(2,5,6;2,3)$

Theorem 8. *GFM*$(2,5,6;2,3) =$ *RE.*

For the next theorem, we again apply the (modified) $(4+1,1)$-GNF.

Theorem 9. *GFM*$(2,3,7;4,2) =$ *RE.*

Instead of giving a full proof here, we only display the matrices simulating $ABC \to \lambda$ in Fig. 4 and show the intended use of these matrices in the following.

$$\begin{aligned} w = \alpha ABC\beta t &\Rightarrow_{m1} \alpha A\$\$\#\#\beta t \Rightarrow_{m2} \alpha\#\$\#\$\#\#\beta t \\ &\Rightarrow_{m3} \alpha\#\$\#\#\beta t \Rightarrow_{m3} \alpha\#\#\beta t \Rightarrow_{m4} \alpha\beta t \end{aligned}$$

Further details are to be found in the long version of this paper. □

We present two more constructions that start out with a $(4+1,1)$-GNF grammar. In each case, we only show the matrices simulating $ABC \to \lambda$ and how this should actually work in a simulation. The formal correctness proofs are contained in the long version of this paper.

Theorem 10. *GFM*$(2,3,8;3,2) =$ *RE.*

The simulation of $ABC \to \lambda$ by the matrices shown in Fig. 5a is as follows:

$$w = \alpha ABC\beta t \Rightarrow_{m1} \alpha A\$\#\beta t \Rightarrow_{m2} \alpha \dagger \$\beta t \Rightarrow_{m3} \alpha\beta t$$

The reader is encouraged to check how the forbidden strings disable unintended derivations. □

Finally, we show how we can bring down both degree and index to two, as well as the matrix length, at the expense of having eight nonterminals and six conditional matrices.

$m1 = [(B \to \$\$), (C \to \#\#), \{S, BB, \#\}]$
$m2 = [(\$ \to \lambda), (A \to \#\$\#), \{\$C, C\$, \#\$\}]$
$m3 = [(\$ \to \lambda), (\# \to \lambda), \{\$A, A\$, \$\$\}]$
$m4 = [(\# \to \lambda), (\# \to \lambda), \{\$\}]$

Fig. 4. Simulating $ABC \to \lambda$, using rules of type $\mathsf{GFM}(2,3,7;4,2)$

$m1 = [(B \to \$), (C \to \#), \{S, BB, \$\}]$
$m2 = [(\# \to \lambda), (A \to \dagger), \{C\$, \$C, \dagger\}]$
$m3 = [(\$ \to \lambda), (\dagger \to \lambda), \{\$A, A\$\}]$

(a) with rules of type $\mathsf{GFM}(2, 3, 8; 3, 2)$

$m1 = [(B \to \$), (C \to C), \{S, \$\}]$
$m2 = [(\$ \to \$), (C \to \#), \{AC, \#\}]$
$m3 = [(\# \to \#\dagger), (\$ \to \lambda), \{B\#, \dagger\}]$
$m4 = [(\dagger \to \lambda), (A \to \$\$\dagger), \{C\#, \$\}]$
$m5 = [(\# \to \lambda), (\dagger \to \lambda), \{\dagger A, \#\dagger\}]$
$m6 = [(\$ \to \lambda), (\$ \to \lambda), \{\dagger, \$B\}]$

(b) with rules of type $\mathsf{GFM}(2, 2, 8; 6, 2)$

Fig. 5. Simulating $ABC \to \lambda$

Theorem 11. $\mathit{GFM}(2, 2, 8; 6, 2) = \mathit{RE}$.

The simulation of $ABC \to \lambda$ by the matrices shown in Fig. 5 is as follows:

$$\begin{aligned} w = \alpha ABC\beta t &\Rightarrow_{m1} \alpha A\$C\beta t \Rightarrow_{m2} \alpha A\$\#\beta t \Rightarrow_{m3} \alpha A\#\dagger\beta t \\ &\Rightarrow_{m4} \alpha\$\$\dagger\#\beta t \Rightarrow_{m5} \alpha\$\$\beta t \Rightarrow_{m6} \alpha\beta t \end{aligned}$$

In this case, proving that unintended derivations are impossible requires some work, as contained in the long version of this paper. □

6 Connection to Membrane Computing

In this section, we explain in detail the connections of generalized forbidding matrix grammars to membrane computing which was already sketched in the introduction in a brief manner. We show how the results obtained in this paper, whenever the maximum length of matrices is 2, correspond to P systems with a membrane structure of depth 2 under the assumption that matrices of length one are processed in the outer (skin) membrane.

To explain the correspondence in a better and more formal way, we recall the definition of an extended rewriting P (ERP) system according to [11] as follows.

Definition 1. *[11] An extended rewriting P system with forbidding rules and of degree $m \geq 1$, denoted as $\mathsf{ERP}_m(forb)$ system, is a construct*

$$\Pi = (V, T, \mu, M_1, \ldots, M_m, R_1, \ldots, R_m)$$

where:

1. *V is the alphabet of the system;*
2. *$T \subseteq V$ is the terminal alphabet;*
3. *μ is a membrane structure with m membranes injectively labeled by $1, \ldots, m$;*
4. *$M_1, \ldots, M_m$ are finite languages over V, representing the strings initially present in the regions $1, 2, \ldots, m$ of the system;*
5. *$R_1, \ldots, R_m$ are finite sets of rules of the form $X \to (u, tar); F$, with $X \in V, u \in V^*, tar \in \{here, out, in\}$, associated with the regions of μ and $F \subseteq V$.*

The rule $X \to x$ can be applied only to the strings which do not contain any symbol from F that is associated with the rule (when $F = \emptyset$ this means that the rule is applied without any restriction, in the free mode). A sequence of transitions forms a computation and the result of a halting computation (where no further transition is possible) is the set of strings over T sent out of the system during the computation. A computation which never halts yields no result. A string which contains symbols not in T does not contribute to the generated language. We denote by $L_{forb}(\Pi)$ the language generated by a P system Π using such rules, and we denote by $\mathsf{ERP}_m(forb)$, $m \geq 1$, the family of all such languages, generated by systems with at most m membranes.

To establish connection between generalized forbidding grammars and ERP system with forbidding rules, we consider a new variant of $\mathsf{ERP}_m(forb)$ called *generalized* ERP *system with forbidding set*, denoted by $\mathsf{GERP}_m(forb)$, where instead set of forbidden symbols (of $\mathsf{ERP}_m(forb)$), we consider a set of forbidden strings and the output is not the environment, rather it is the skin membrane. We now define the new variant formally. A generalized ERP system with forbidding sets Π' having $m \geq 1$ membranes, n nonterminals, degree d and index i is a construct

$$(V, T, \mu, M_1, \ldots, M_m, R_1, \ldots, R_m; 1)$$

where:

- V is the alphabet; $T \subseteq V$ is terminal alphabet; with $n = |V \setminus T|$
- μ is a membrane structure $[_1\ [_2\ ]_2\ [_3\ ]_3 \cdots [_m\ ]_m]\ _1$ with m membranes where membrane 1 is the skin membrane as well as the output membrane;
- $M_1, \ldots, M_m$ are strings (over V) in regions $1, 2, \ldots, m$;
- $R_1, \ldots, R_m$ are finite set of rules present in regions $1, 2, \ldots, m$;
- Rule form in R_k: $[X \to (x, tar); F_k]$ where $1 \leq k \leq m$ and
  - $X \in V \setminus T$; $x \in V^*$; $tar \in \{in_j, out, here\}$; $2 \leq j \leq m$.
  - $F_k \subseteq V^*$ (with strings of length $\leq d$) with $|F_k| \leq i$ for all $1 \leq k \leq m$.

We denote by $L_{Gforb}(\Pi')$ the language generated by a P system Π' using such rules, and we denote by $\mathsf{GERP}_m(forb_d; i, n)$, $m, n \geq 1$ and $d, n \geq 0$, the family of all such languages, generated by systems with m membranes, n nonterminals, degree d and index i.

Having defined $\mathsf{GERP}_m(forb_d; i, n)$, we are now ready to establish a relationship between GFM and GERP. If a matrix m of a GFM system with length two is chosen for application, then the first rule of the matrix is applied in the skin membrane labelled as 1, thus implying that all the first rules of matrices are present in skin membrane. The second rule of the matrix m is applied in some inner membrane (say, membrane $j > 1$). From the inner membrane, the resultant string goes back to the skin membrane. Every time before a rule m is applied in the skin membrane, the forbidding set corresponding to m acts as a filtration check to move to the inner membrane specified in m. This helps obtain computational completeness results for GERP systems with forbidding sets having bounded resources in nonterminals and rewriting rules.

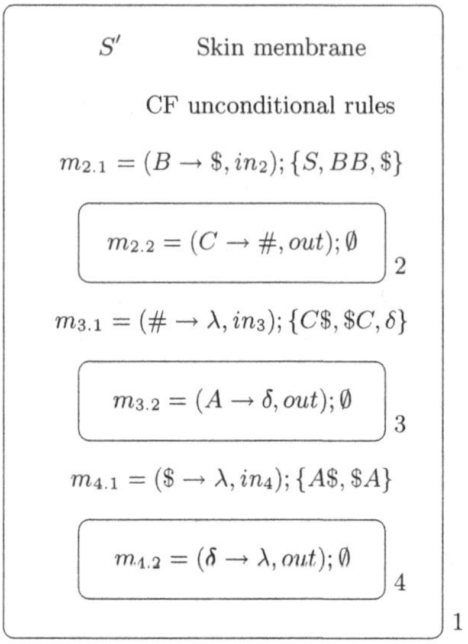

(a) $\mathsf{GERP}_4(forb_2; 3, 8)$ built from:

$[(S' \to uS'a), \emptyset]$
$[(S' \to uSv), \emptyset]$, $[(S' \to uv), \emptyset]$
$[(S \to uSv), \emptyset]$, $[(S \to uv), \emptyset]$
$m2 = [(B \to \$), (C \to \#), \{S, BB, \$\}]$
$m3 = [(\# \to \lambda), (A \to \delta), \{C\$, \$C, \delta\}]$
$m4 = [(\$ \to \lambda), (\delta \to \lambda), \{\$A, A\$\}]$

(b) $\mathsf{GFM}(2, 3, 8; 3, 2)$

Fig. 6. Correspondence between $\mathsf{GERP}_4(forb_2; 3, 8)$ and $\mathsf{GFM}(2, 3, 8; 3, 2)$

We relate the GFM grammar $G = (V, T, M, S)$ corresponding to the language family $\mathsf{GFM}(d, i, n; m-1, 2)$, where the $m-1$ matrices are denoted by $m_p = [m_{p,1}, m_{p,2}, F_p](2 \le p \le m)$, with the generalized ERP system Π', coming from $\mathsf{GERP}_m(forb_d; i, n)$, as follows:

- $\Pi' = (V, T, \mu, M_1, \ldots, M_m, R_1, \ldots, R_m; 1)$;
- $\mu = [_1\ [_2\ ]_2\ [_3\ ]_3 \cdots [_m\ ]_m\ ]_1$;
- $M_1 = S$, $M_k = \emptyset$ for all $2 \le k \le m$;
- $R_1 = C \cup D$, where
  - $C = \{X \to (x, here); \emptyset \mid [(X \to x), \emptyset] \in M\}$
  - $D = \bigcup_{p=2}^{m} \{(m_{p,1}, in_p); F_p\}$ where $m_{p,1}$ is the first rule of matrix $m_p \in M$.
- For all $2 \le k \le m$, $R_k = (m_{k,2}, out); \emptyset$. Informally, the inner membrane k contains only one rule R_k which is the second rule in matrix m_k of G.

Figure 6 illustrates the correspondence between $\mathsf{GFM}(2, 3, 8; 3, 2)$ and the generalized ERP system $\mathsf{GERP}_4(forb_2; 3, 8)$ as discussed above.

With this correspondence, our results of this paper correspond to the following results for GERP systems (for the first case in the list, we refer to [11]).

Corollary 2. *The following GERP systems are computational complete:* $\mathsf{GERP}_2(forb_1; *, *)$, $\mathsf{GERP}_*(forb_1; 1, *)$, $\mathsf{GERP}_4(forb_2; 5, 6)$, $\mathsf{GERP}_4(forb_2; 3, 8)$, $\mathsf{GERP}_5(forb_2; 3, 7)$, $\mathsf{GERP}_7(forb_2; 2, 8)$. □

7 Discussions

We have introduced generalized forbidding matrix grammars as a further variant of regulated rewriting grammar systems and established connections with membrane computing. Let us point to several open problems in this area.

- We have proven that GFM grammars of degree one and index one are computationally complete, but we do not have such results if all (other) parameters are also bounded by some constant. This is also the case for larger values of degree or index.
- We have shown that three nonterminals are sufficient for arriving at computational completeness in general for GFM grammars (Theorem 1), but again then most other parameters are unbounded. This should be compared to the best we could do for GFM grammars with bounded parameters, which is getting down to six nonterminals, the index (five) being relatively high.
- It would be also interesting to obtain computational incompleteness results. For instance, are there any recursively enumerable languages that cannot be generated by a GFM grammar with two nonterminals? More such open problems show up if all parameters are bounded.
- Whether the results and ideas used in this paper or in [7] are useful to improve the computational completeness results of P systems with inhibitors is left open, connecting to multiset rewriting.
- As can be seen in Fig. 6, the inner membranes have no filtration tests attached to them. We could change this by allowing (in the original matrix grammar model) tests attached to single rules, not to matrices. This would give some additional flexibility, so it would be interesting to compare such grammars with the GFM grammars studied here.
- In the context of insertion-deletion systems, we described the interpretation of certain graph-control mechanisms (similar to matrix grammars) to P systems to some detail in [9]. Hence, similar ideas, interpreting generalized forbidding insertion-deletion systems [10] to P systems, could be applied here.
- As a final remark, let us mention that we are not aware on any studies on the nonterminal complexity of P systems; only the work of Madhu [16] goes in this direction. This would be an interesting topic of future research.

Acknowledgement. Without the numerous contributions of Gheorghe Păun to the theory of Formal Languages, the present paper could hardly be written, as it is based in particular on [12,22,23]. Also, the second author profusely thanks Gheorghe Păun for being his source of inspiration since from his Ph.D. days. Happy birthday, Gheorghe!

References

1. Ábrahám, S.: Some questions of phrase-structure grammars. I. Comput. Linguist. **4**, 61–70 (1965)
2. Bottoni, P., Labella, A., Martín-Vide, C., PĂun, G.: Rewriting P systems with conditional communication. In: Brauer, W., Ehrig, H., Karhumäki, J., Salomaa, A. (eds.) Formal and Natural Computing. LNCS, vol. 2300, pp. 325–353. Springer, Heidelberg (2002). https://doi.org/10.1007/3-540-45711-9_18

3. Bottoni, P., Martín-Vide, C., Păun, Gh, Rozenberg, G.: Membrane systems with promoters/inhibitors. Acta Informatica **38**(10), 695–720 (2002)
4. Dassow, J., Păun, Gh: Regulated Rewriting in Formal Language Theory, EATCS Monographs in Theoretical Computer Science, vol. 18. Springer (1989)
5. Fernau, H.: Nonterminal complexity of programmed grammars. Theor. Comput. Sci. **296**, 225–251 (2003)
6. Fernau, H., Freund, R., Oswald, M., Reinhardt, K.: Refining the nonterminal complexity of graph-controlled, programmed, and matrix grammars. J. Automata Lang. Comb. **12**(1/2), 117–138 (2007)
7. Fernau, H., Kuppusamy, L., Oladele, R.O., Raman, I.: Improved descriptional complexity results on generalized forbidding grammars. In: Pal, S.P., Vijayakumar, A. (eds.) CALDAM 2019. LNCS, vol. 11394, pp. 174–188. Springer, Cham (2019). https://doi.org/10.1007/978-3-030-11509-8_15
8. Fernau, H., Kuppusamy, L., Oladele, R.O., Raman, I.: Improved descriptional complexity results on generalized forbidding grammars. Disc. Appl. Math. (2021). https://doi.org/10.1016/j.dam.2020.12.027
9. Fernau, H., Kuppusamy, L., Raman, I.: On path-controlled insertion-deletion systems. Acta Informatica **56**(1), 35–59 (2019)
10. Fernau, H., Kuppusamy, L., Raman, I.: On the power of generalized forbidding insertion-deletion systems. In: Jirásková, G., Pighizzini, G. (eds.) DCFS 2020. LNCS, vol. 12442, pp. 52–63. Springer, Cham (2020). https://doi.org/10.1007/978-3-030-62536-8_5
11. Ferretti, C., Mauri, G., Paun, Gh, Zandron, C.: On three variants of rewriting P systems. Theor. Comput. Sci. **1–3**(301), 201–215 (2003)
12. Freund, R., Păun, G.: On the number of non-terminal symbols in graph-controlled, programmed and matrix grammars. In: Margenstern, M., Rogozhin, Y. (eds.) MCU 2001. LNCS, vol. 2055, pp. 214–225. Springer, Heidelberg (2001). https://doi.org/10.1007/3-540-45132-3_14
13. Geffert, V.: Normal forms for phrase-structure grammars. RAIRO Informatique théorique et Appl./Theor. Inform. Appl. **25**, 473–498 (1991)
14. Hauschildt, D., Jantzen, M.: Petri net algorithms in the theory of matrix grammars. Acta Informatica **31**, 719–728 (1994)
15. Ionescu, M., Sburlan, D.: On P systems with promoters/inhibitors. J. Universal Comput. Sci. **10**(5), 581–599 (2004)
16. Madhu, M.: Descriptional complexity of rewriting P systems. J. Automata, Lang. Comb. **9**(2–3), 311–316 (2004)
17. Masopust, T., Meduna, A.: Descriptional complexity of generalized forbidding grammars. In: Geffert, V., Pighizzini, G. (eds.) 9th International Workshop on Descriptional Complexity of Formal Systems - DCFS, pp. 170–177. University of Kosice, Slovakia (2007)
18. Masopust, T., Meduna, A.: Descriptional complexity of grammars regulated by context conditions. In: Loos, R., Fazekas, S.Z., Martín-Vide, C. (eds.) LATA 2007. Proceedings of the 1st International Conference on Language and Automata Theory and Applications. vol. Report 35/07, pp. 403–412. Research Group on Mathematical Linguistics, Universitat Rovira i Virgili, Tarragona (2007)
19. Meduna, A.: Generalized forbidding grammars. Int. J. Comput. Math. **36**, 31–39 (1990)
20. Meduna, A., Svec, M.: Descriptional complexity of generalized forbidding grammars. Int. J. Comput. Math. **80**(1), 11–17 (2003)

21. Păun, Gh: Six nonterminals are enough for generating each r.e. language by a matrix grammar. Int. J. Comput. Math. **15**(1–4), 23–37 (1984)
22. Păun, Gh: A variant of random context grammars: semi-conditional grammars. Theor. Comput. Sci. **41**, 1–17 (1985)
23. Păun, Gh: Computing with membranes. J. Comput. Syst. Sci. **61**(1), 108–143 (2000)

Parallel Contextual Array Insertion Deletion P Systems and Tabled Matrix Grammars

S. James Immanuel[1], S. Jayasankar[2(✉)], D. Gnanaraj Thomas[3], and Meenakshi Paramasivan[4]

[1] Department of Mathematics, Sri Sairam Institute of Technology, Chennai 600044, India

[2] Department of Mathematics, Ramakrishna Mission Vivekananda College, Chennai 600004, India

[3] Department of Applied Mathematics, Saveetha School of Engineering, SIMATS, Chennai 602105, India

[4] FB IV - Informatikwissenschaften, Universität Trier, 54286 Trier, Germany

Abstract. Siromoney et al. introduced a parallel/sequential generative model called Tabled Matrix Grammars (TMGs) by generalising phrase structure matrix grammars generating abstract families of languages (AFLs). James et al. introduced Parallel Contextual Array Insertion Deletion P Systems (PCAIDPSs) to generate two-dimensional array languages using insertion and deletion operations through parallel contextual mappings. In this paper, we compare the generative powers of PCAIDPSs and TMGs. We prove that the family of languages generated by PCAIDPS with two membranes properly includes the family of languages generated by Tabled Context-sensitive Matrix Grammars (TCSMGs).

Keywords: Parallel contextual array grammars · Tabled matrix grammars · Insertion deletion P systems

1 Introduction

Siromoney et al. [30] introduced a parallel/ sequential grammar model called tabled matrix model more general than Phrase Structure Matrix Grammar (PSMG) [29]. In this model, the production rules for the vertical generation were given in the form of tables comprising of a finite set of right linear rules. They studied closure properties and compared with other array generating models with respect to their generative capacities. The effect of control on the tables was studied and it was shown that the regular control does not increase the generative capacity while context-free (CF) and context-sensitive (CS) controls do. Different kinds of control may be imposed on any type of grammar. The effects of "string control" and "array control" on the matrix models were studied in [24]. Still yet a third kind of control is introduced in [28] for parallel grammars

R. Freund et al. (Eds.): CMC 2020, LNCS 12687, pp. 46–77, 2021.
https://doi.org/10.1007/978-3-030-77102-7_4

and Siromoney et al. [30] found that this kind of control imposed on the matrix models [29] yielded the tabled matrix models which generalized the results of [28].

In 1969 Marcus [25] came out with an entirely different class of grammars different from Chomsky grammars called contextual grammars. A contextual grammar produces a language by starting from a given finite set of strings and adding repetitively, pairs of strings (called as contexts), associated to sets of words (called selectors) to the string already obtained. Both Freund et al and Helen Chandra et al. [4,9] extended these grammars to two-dimensional arrays respectively in their own style. Freund et al generalised the concept of contextual grammars and adopted a new and simple approach in [9]. Both row and column contexts are allowed and contextual rules are finite in parallel contextual array grammars [4].

D. Haussler [12] was the first to conceive context-free insertion systems as a generalisation of concatenation. L. Kari [20] studied the role of insertion and deletion operations in formal language theory in 1991. Domaratzki and Okhotin [7] and Ito, Kari, Thierrin and Yu [18,21] investigated different variations of insertion and deletion systems.

A P system (membrane system) introduced by Păun [26] is a distributed theoretical model with maximal parallelism based on the membrane structure and function of the living cells. In a P system, at each step, all objects which can evolve should evolve in parallel. Computation of a P system starts from its initial configuration guided by the structure of a membrane, objects it contains and evolution rules of that membrane and terminates when the rules of the membrane are exhausted. P systems have proved to be a rich theoretical framework to study many computational problems besides giving a new impetus to formal language theory. Various types of P systems were introduced in the literature and their properties, computing power, normal forms and basic decision problems were studied [26,27]. S. N. Krishna et al. [22] introduced a new variant of P system with string objects having insertion-deletion rules to control the production of strings. A. Alhazov [1] et al have considered insertion-deletion P systems with priority over insertion and showed that these P systems with one-symbol together with context-free insertion rules were able to generate Parikh sets of all recursively enumerable languages (PsRE). The contextual way of handling string objects in P systems has been considered by Madhu et al. [23] and the contextual P systems are found to be more powerful than ordinary string contextual grammars and their variants. Ceterchi et al. [2] introduced array P systems of the isometric variety, extending the string rewriting P systems to arrays using context-free type of rules. In [8], a P system model called contextual array P system with array objects and array contextual rules has been introduced based on the contextual style of array generation considered in [9], and its generative power in the description of picture arrays was examined. In [5], P system models namely, external and internal array contextual P systems were introduced. Influenced by the works on contextual style of external and internal parallel contextual array grammars [4,31], James et al. [17] introduced a new P system model called external and internal parallel contextual array P systems and parallel contextual array P systems by shuffling contexts on trajectories and studied some properties of the family of languages generated by these P systems and compared their generative capacity with other array generating P systems. James et al also have introduced another new contextual array P system subsequently called parallel contextual array insertion deletion P system (PCAIDPS) in [15], based on internal parallel contextual array grammars in [4] and proved that the families of local (LOC) and recognizable picture languages (REC) [10,11] of Giammaresi and Resitivo and context-sensitive matrix languages ($CSML$) [29] of Siromoney et al are properly contained in the family of languages generated by the parallel contextual array insertion deletion P systems with 2 membranes ($\mathscr{L}(PCAIDPS_2)$) [15]. Jayasankar et al [13,19] proved that $\mathscr{L}(PCAIDPS_2)$ properly contains the family of array languages generated by (Context-free : Right-linear Indexed Right-linear) Siromoney matrix grammars ($\mathscr{L}(CF : RIR)$).

In this paper, we take a step forward to find a hierarchy among the two-dimensional picture languages by comparing the generative powers of the models Parallel Contextual Array Insertion Deletion Grammars (PCAIDGs) [16], PCAIDPSs [15] with that of Tabled Matrix Grammars (TMGs) [30]. We show that $\mathscr{L}(TCSMG) \subsetneq \mathscr{L}(PCAIDPS_2)$.

The paper is organised as follows: In Sect. 2, some basic definitions pertaining to Tabled Matrix Grammars and languages with examples are recalled. In Sect. 3, definitions of PCAIDGs and PCAIDPSs are given along with interesting examples. In Sect. 4, main results: $\mathscr{L}(TCSMG) \subsetneq \mathscr{L}(PCAIDG)$ and $\mathscr{L}(TCSMG) \subsetneq \mathscr{L}(PCAIDPS_2)$ are proved. In Sect. 5, some open problems are suggested along with the conclusion.

2 Preliminaries

In this section, we recall some definitions pertaining to tabled matrix grammar (TMG) as in [30].

Let V be a finite alphabet, V^*, the set of words over V including the empty word λ. $V^+ = V^* - \{\lambda\}$. An array consists of finitely many symbols from V arranged in rows and columns and is written in the form, $A = \begin{bmatrix} a_{11} & \cdots & a_{1n} \\ \vdots & \ddots & \vdots \\ a_{m1} & \cdots & a_{mn} \end{bmatrix}$ or $\begin{matrix} a_{11} & \cdots & a_{1n} \\ \vdots & \ddots & \vdots \\ a_{m1} & \cdots & a_{mn} \end{matrix}$ or in short $A = [a_{ij}]_{m\times n}$, $a_{ij} \in V, i = 1, 2, \ldots, m$ and $j = 1, 2, \ldots, n$. The set of all arrays over V is denoted by V^{**} which also includes the empty array Λ (zero rows and zero columns). $V^{++} = V^{**} - \{\Lambda\}$. The column concatenation of $A = \begin{bmatrix} a_{11} & \cdots & a_{1p} \\ \vdots & \ddots & \vdots \\ a_{m1} & \cdots & a_{mp} \end{bmatrix}$ and $B = \begin{bmatrix} b_{11} & \cdots & b_{1q} \\ \vdots & \ddots & \vdots \\ b_{n1} & \cdots & b_{nq} \end{bmatrix}$, defined only when $m = n$, is given by $A \oplus B = \begin{bmatrix} a_{11} & \cdots & a_{1p} & b_{11} & \cdots & b_{1q} \\ \vdots & \ddots & \vdots & \vdots & \ddots & \vdots \\ a_{m1} & \cdots & a_{mp} & b_{n1} & \cdots & b_{nq} \end{bmatrix}$.

As $1 \times n$ arrays can be easily interpreted as words of length n (and vice versa), we will then write their column concatenation by juxtaposition (as usual). Similarly, the row concatenation of A and B, defined only when $p = q$, is given by $A \ominus B = \begin{bmatrix} a_{11} & \cdots & a_{1p} \\ \vdots & \ddots & \vdots \\ a_{m1} & \cdots & a_{mp} \\ b_{11} & \cdots & b_{1q} \\ \vdots & \ddots & \vdots \\ b_{n1} & \cdots & b_{nq} \end{bmatrix}$. The empty array acts as the identity for column and row concatenation of arrays of arbitrary dimensions.

Definition 1. *A* Tabled Context-sensitive Matrix Grammar (TCSMG) (Tabled Context-free Matrix Grammar (TCFMG), Tabled Regular Matrix Grammar (TRMG)) is a two-tuple $G = (G_H, G_V)$ where $G_H = (N_H, I, P_H, S)$ is a Context-sensitive Grammar (CSG) (Context-free Grammar (CFG), Regular Grammar (RG)) *with*

- N_H *is a finite set of horizontal non-terminals;*
- $I = \{S_1, \ldots, S_k\}$ *is a finite set of intermediates and* $N_H \cap I = \emptyset$*;*
- P_H *is a finite set of context-sensitive horizontal rules;*
- $S \in N_H$ *is the start symbol; and*
- $G_V = \left(\bigcup_{i=1}^{k} G_i, \mathscr{P}\right)$ *where each* $G_i = (N_i, T, P_i, S_i)$ *is a right-linear grammar with*
  - N_i *is a finite set of vertical non-terminals;* $N_i \cap N_j = \emptyset$ *if* $i \neq j$*;*
  - T *is a finite set of terminals;*
  - $P_i = P_{N_i} \cup P_{T_i}$*;*
  - P_{N_i} *is a finite set of right-linear non-terminal production rules of the form* $A \to aB$*;*
  - P_{T_i} *is a finite set of right-linear terminal production rules of the form* $A \to a$*, where* $A, B \in N_i$*,* $a \in T \cup \{\lambda\}$*;*
  - $S_i \in N_i$ *is the start symbol of* G_i*;*
  - $\mathscr{P}$ *is a finite set of tables containing production rules.*

Each non-terminal table $t \in \mathscr{P}$ is a non-empty subset of $\bigcup_{i=1}^{k} P_{N_i}$ and each terminal table $t \in \mathscr{P}$ is a non-empty subset of $\bigcup_{i=1}^{k} P_{T_i}$. The rules of each grammar G_i are defined in such a way that the rectangular structure is preserved during vertical derivation.

The derivations in a TCSMG proceed as follows: First a string $S_{i_1} \ldots S_{i_n} \in I^*$, $i_j \in \{1, \ldots k\}$ is generated by the horizontal rules in P_H. Vertical derivations take place in parallel restricted by tables of $\mathscr{P}$. We write $M_1 \Downarrow_{G_V} M_2$ if and only if either

i) $M_1 = \begin{array}{ccc} a_{11} & \dots & a_{1n} \\ \vdots & \ddots & \vdots \\ a_{(m-1)1} & \dots & a_{(m-1)n} \\ A_1 & \dots & A_n \end{array}, \quad M_2 = \begin{array}{ccc} a_{11} & \dots & a_{1n} \\ \vdots & \ddots & \vdots \\ a_{(m-1)1} & \dots & a_{(m-1)n} \\ a_{m1} & \dots & a_{mn} \\ B_1 & \dots & B_n \end{array}$

and t is a non-terminal table of $\mathscr{P}$ such that $A_i \rightarrow a_{mi}B_i$, $i = 1, \dots, n$ are rules in t or

ii) $M_1 = \begin{array}{ccc} a_{11} & \dots & a_{1n} \\ \vdots & \ddots & \vdots \\ a_{(m-1)1} & \dots & a_{(m-1)n} \\ A_1 & \dots & A_n \end{array}, \quad M_2 = \begin{array}{ccc} a_{11} & \dots & a_{1n} \\ \vdots & \ddots & \vdots \\ a_{(m-1)1} & \dots & a_{(m-1)n} \\ u_{m1} & \dots & a_{mn} \end{array}$

and t is a terminal table of $\mathscr{P}$ such that $A_i \rightarrow a_{mi}$, $i = 1, \dots, n$ are rules in t. $\Downarrow^*$ is the transitive closure of $\Downarrow$.

Definition 2. *The set of all matrices generated by G is defined as,*

$$L(G) = \left\{ [a_{ij}] \mid S \Rightarrow^*_{G_H} S_{i_1} \dots S_{i_n} \Downarrow^*_{G_V} [a_{ij}] \right\}.$$

$L(G)$ is called as a Tabled Context-sensitive Matrix Language (TCSML), Tabled Context-free Matrix Language (TCFML), Tabled Regular Matrix Language (TRML), if G is a TCSMG, TCFMG, TRMG respectively. The corresponding families of languages are denoted by $\mathscr{L}(TCSMG)$, $\mathscr{L}(TCFMG)$ and $\mathscr{L}(TRMG)$ respectively.

Example 1. Let us consider the language $L_1 \in \mathscr{L}(TRMG)$ consisting of arrays describing H tokens of different sizes with varied proportions i.e.,

$$L_1 = \left\{ \begin{array}{ccc} x & \bullet & x \\ x & x & x \\ x & \bullet & x \end{array}, \begin{array}{ccc} x & \bullet & x \\ x & \bullet & x \\ x & x & x \\ x & \bullet & x \end{array}, \begin{array}{ccc} x & \bullet & x \\ x & x & x \\ x & \bullet & x \\ x & \bullet & x \end{array}, \begin{array}{cccc} x & \bullet & \bullet & x \\ x & x & x & x \\ x & \bullet & \bullet & x \end{array}, \begin{array}{ccc} x & \bullet & x \\ x & \bullet & x \\ x & x & x \\ x & \bullet & x \\ x & \bullet & x \end{array}, \begin{array}{ccc} x & \bullet & x \\ x & x & x \\ x & \bullet & x \\ x & \bullet & x \\ x & \bullet & x \end{array}, \begin{array}{ccc} x & \bullet & x \\ x & \bullet & x \\ x & \bullet & x \\ x & x & x \\ x & \bullet & x \end{array}, \cdots \right\}.$$

L_1 is generated by the TRMG $G_1 = (G_H, G_V)$ where $G_H = (N_H, I, P_H, S)$ with,

- $N_H = \{S, A, B\}$,
- $I = \{S_1, S_2\}$,
- $P_H = \{S \rightarrow S_1A, A \rightarrow S_2B, B \rightarrow S_2B, B \rightarrow S_1\}$.

The language generated by G_H with intermediates $\{S_1, S_2\}$ is

$$L(G_H) = \{S_1S_2^nS_1 \mid n \geq 1\}.$$

The tables of G_V are $t_1 = \{S_1 \rightarrow xS_1, S_2 \rightarrow \bullet S_2\}$, $t_2 = \{S_1 \rightarrow xS_1, S_2 \rightarrow \bullet A\}$, $t_3 = \{S_1 \rightarrow xS_1, A \rightarrow xB\}$, $t_4 = \{S_1 \rightarrow xS_1, B \rightarrow \bullet B\}$, $t_5 = \{S_1 \rightarrow x, B \rightarrow \bullet\}$.

A sample derivation of a picture of size 7×5 in L_1 is given below:

$$S \Rightarrow^*_{G_H} S_1 S_2^3 S_1 = S_1 S_2 S_2 S_2 S_1 \Rightarrow_{t_1} \begin{matrix} x & \bullet & \bullet & \bullet & x \\ S_1 & S_2 & S_2 & S_2 & S_1 \end{matrix} \Rightarrow_{t_1}$$

$$\begin{matrix} x & \bullet & \bullet & \bullet & x \\ x & \bullet & \bullet & \bullet & x \\ S_1 & S_2 & S_2 & S_2 & S_1 \end{matrix} \Rightarrow_{t_2} \begin{matrix} x & \bullet & \bullet & \bullet & x \\ x & \bullet & \bullet & \bullet & x \\ x & \bullet & \bullet & \bullet & x \\ S_1 & A & A & A & S_1 \end{matrix} \Rightarrow_{t_3} \begin{matrix} x & \bullet & \bullet & \bullet & x \\ x & \bullet & \bullet & \bullet & x \\ x & \bullet & \bullet & \bullet & x \\ x & x & x & x & x \\ S_1 & B & B & B & S_1 \end{matrix} \Rightarrow_{t_4}$$

$$\begin{matrix} x & \bullet & \bullet & \bullet & x \\ x & \bullet & \bullet & \bullet & x \\ x & \bullet & \bullet & \bullet & x \\ x & x & x & x & x \\ x & \bullet & \bullet & \bullet & x \\ S_1 & B & B & B & S_1 \end{matrix} \Rightarrow_{t_4} \begin{matrix} x & \bullet & \bullet & \bullet & x \\ x & \bullet & \bullet & \bullet & x \\ x & \bullet & \bullet & \bullet & x \\ x & x & x & x & x \\ x & \bullet & \bullet & \bullet & x \\ x & \bullet & \bullet & \bullet & x \\ S_1 & B & B & B & S_1 \end{matrix} \Rightarrow_{t_5} \begin{matrix} x & \bullet & \bullet & \bullet & x \\ x & \bullet & \bullet & \bullet & x \\ x & \bullet & \bullet & \bullet & x \\ x & x & x & x & x \\ x & \bullet & \bullet & \bullet & x \\ x & \bullet & \bullet & \bullet & x \\ x & \bullet & \bullet & \bullet & x \end{matrix}.$$

Example 2. Consider the TCSMG $G_2 = (G_H, G_V)$, where $G_H = (N_H, I, P_H, S)$ with

- $N_H = \{S, C\}$,
- $I = \{S_1, S_2\}$,
- $P_H = \{S \to S_1 S C S_1, S \to S_1 S_2 S_1 S_2 S_1,\ S_1 C \to C S_1,\ S_2 C \to S_1 S_2\}$.

Two sample derivations of G_H are given below:

$$S \Rightarrow S_1 S C S_1 \Rightarrow S_1(S_1 S_2 S_1 S_2 S_1) C S_1 =$$
$$S_1 S_1 S_2 S_1 S_2 (S_1 C) S_1 \Rightarrow S_1^2 S_2 S_1 S_2 (C S_1) S_1 = S_1^2 S_2 S_1 (S_2 C) S_1^2 \Rightarrow$$
$$S_1^2 S_2 S_1 (S_1 S_2) S_1^2 = S_1^2 S_2 S_1^2 S_2 S_1^2.$$

$$S \Rightarrow S_1 S C S_1 \Rightarrow S_1(S_1 S C S_1) C S_1 \Rightarrow S_1 S_1 (S_1 S_2 S_1 S_2 S_1) C S_1 C S_1 \Rightarrow$$
$$S_1^3 S_2 S_1 S_2 S_1 C (C S_1) S_1 \Rightarrow S_1^3 S_2 S_1 S_2 (C S_1) C S_1^2 \Rightarrow$$
$$S_1^3 S_2 S_1 S_2 C (C S_1) S_1^2 \Rightarrow S_1^3 S_2 S_1 (S_1 S_2) C S_1^3 \Rightarrow S_1^3 S_2 S_1^2 (S_1 S_2) S_1^3 =$$
$$S_1^3 S_2 S_1^3 S_2 S_1^3.$$

The language generated by G_H is

$$L(G_H) = \{S_1^n S_2 S_1^n S_2 S_1^n \mid n \geq 1\}.$$

The tables of G_V are the same as in Example 1. $L(G_2)$ is the set of all matrices over $\{x, \bullet\}$ with following properties. Each matrix in $L(G_2)$

(i) contains one row full of $x's$.
(ii) has $3n + 2, n \geq 1$ columns.
(iii) has $3n$ identical columns with each column full of $x's$.
(iv) has two identical columns having $x's$ neither at the top nor at the bottom of the column.
(v) has m rows with $m \geq 3$.

$L(G_2)$ is a TCSML but not a CSML [30]. Few sample derivations are as follows:

$$S_1S_2S_1S_2S_1 \Rightarrow_{t_2} \begin{array}{ccccc} x & \bullet & x & \bullet & x \\ S_1 & A & S_1 & A & S_1 \end{array} \Rightarrow_{t_3} \begin{array}{ccccc} x & \bullet & x & \bullet & x \\ x & x & x & x & x \\ S_1 & B & S_1 & B & S_1 \end{array} \Rightarrow_{t_5} \begin{array}{ccccc} x & \bullet & x & \bullet & x \\ x & x & x & x & x \\ x & \bullet & x & \bullet & x \end{array}.$$

$$S_1S_2S_1S_2S_1 \Rightarrow_{t_2} \begin{array}{ccccc} x & \bullet & x & \bullet & x \\ S_1 & A & S_1 & A & S_1 \end{array} \Rightarrow_{t_3} \begin{array}{ccccc} x & \bullet & x & \bullet & x \\ x & x & x & x & x \\ S_1 & B & S_1 & B & S_1 \end{array} \Rightarrow_{t_4}$$

$$\begin{array}{ccccc} x & \bullet & x & \bullet & x \\ x & x & x & x & x \\ x & \bullet & x & \bullet & x \\ S_1 & B & S_1 & B & S_1 \end{array} \Rightarrow_{t_5} \begin{array}{ccccc} x & \bullet & x & \bullet & x \\ x & x & x & x & x \\ x & \bullet & x & \bullet & x \\ x & \bullet & x & \bullet & x \end{array}.$$

$$S_1^2S_2S_1^2S_2S_1^2 \Rightarrow_{t_2} \begin{array}{cccccccc} x & x & \bullet & x & x & \bullet & x & x \\ S_1 & S_1 & A & S_1 & S_1 & A & S_1 & S_1 \end{array} \Rightarrow_{t_3}$$

$$\begin{array}{cccccccc} x & x & \bullet & x & x & \bullet & x & x \\ x & x & x & x & x & x & x & x \\ S_1 & S_1 & B & S_1 & S_1 & B & S_1 & S_1 \end{array} \Rightarrow_{t_5} \begin{array}{cccccccc} x & x & \bullet & x & x & \bullet & x & x \\ x & x & x & x & x & x & x & x \\ x & x & \bullet & x & x & \bullet & x & x \end{array}.$$

3 Parallel Contextual Array Insertion Deletion Grammar and Associated P System

In this section, we recall some notions of parallel contextual array insertion deletion grammar and associated P system. For further details we can refer to [15,16].

Definition 3. *Let V be a finite alphabet. A* column array context *over V is of the form* $c = \begin{bmatrix} u_1 \\ u_2 \end{bmatrix} \in V^{**}$, *$u_1, u_2$ are of size $1 \times p$, $p \geq 1$. A* row array context *over V is of the form* $r = \begin{bmatrix} u_1 \; u_2 \end{bmatrix} \in V^{**}$, *$u_1, u_2$ are of size $p \times 1$, $p \geq 1$.*

Definition 4. *The* parallel column contextual insertion operation *is defined as follows: Let V be an alphabet, C be a finite subset of V^{**} whose elements are the column array contexts and $\varphi_c^i : V^{**} \times V^{**} \to 2^C$ be a choice mapping.*

We define $\varphi_c^i : V^{**} \times V^{**} \to 2^{V^{**}}$ such that, for arrays

$$A = \begin{bmatrix} a_{1j} & \cdots & a_{1(k-1)} \\ \vdots & \ddots & \vdots \\ a_{mj} & \cdots & a_{m(k-1)} \end{bmatrix}, B = \begin{bmatrix} a_{1k} & \cdots & a_{1(l-1)} \\ \vdots & \ddots & \vdots \\ a_{mk} & \cdots & a_{m(l-1)} \end{bmatrix}, j < k < l, a_{ij} \in V,$$

$$I_c \in \varphi_c^i(A, B), I_c = \begin{bmatrix} u_1 \\ u_2 \\ \vdots \\ u_m \end{bmatrix},$$

if $c_i = \begin{bmatrix} u_i \\ u_{i+1} \end{bmatrix} \in \varphi_c^i \left(\begin{matrix} a_{ij} & \cdots & a_{i(k-1)} \\ a_{(i+1)j} & \cdots & a_{(i+1)(k-1)} \end{matrix}, \begin{matrix} a_{ik} & \cdots & a_{i(l-1)} \\ a_{(i+1)k} & \cdots & a_{(i+1)(l-1)} \end{matrix} \right)$,
$c_i \in C$, $1 \leq i \leq m-1$, *not all need to be distinct.*

Given an array $X = [a_{ij}]_{m \times n}$, $a_{ij} \in V$, $X = X_1 \oplus A \oplus B \oplus X_2$ with

$$X_1 = \begin{bmatrix} a_{11} & \cdots & a_{1(j-1)} \\ \vdots & \ddots & \vdots \\ a_{m1} & \cdots & a_{m(j-1)} \end{bmatrix}, A = \begin{bmatrix} a_{1j} & \cdots & a_{1(k-1)} \\ \vdots & \ddots & \vdots \\ a_{mj} & \cdots & a_{m(k-1)} \end{bmatrix},$$

$$B = \begin{bmatrix} a_{1k} & \cdots & a_{1(l-1)} \\ \vdots & \ddots & \vdots \\ a_{mk} & \cdots & a_{m(l-1)} \end{bmatrix}, X_2 = \begin{bmatrix} a_{1l} & \cdots & a_{1n} \\ \vdots & \ddots & \vdots \\ a_{ml} & \cdots & a_{mn} \end{bmatrix},$$

$1 \leq j \leq k < l \leq n+1$ *(or)* $1 \leq j < k \leq l \leq n+1$, *we write* $X \Rightarrow^{col_i} Y$ *if* $Y = X_1 \oplus A \oplus I_c \oplus B \oplus X_2$, *such that* $I_c \in \varphi_c^i(A, B)$. *$I_c$ is called the inserted column context. We say that Y is obtained from X by parallel column contextual insertion operation. Note that φ_c^i is column insertion partial function.*

The following 4 special cases for $X = X_1 \oplus A \oplus B \oplus X_2$ are also considered,

1. *For $j = 1$, we have $X_1 = \Lambda$.*
2. *For $j = k$, we have $A = \Lambda$. If $j = k = 1$ then $X_1 = \Lambda$ and $A = \Lambda$.*
3. *For $k = l$ (for $k + p = l$), we have $B = \Lambda$.*
4. *For $l = n+1$, we have $X_2 = \Lambda$. If $k = l = n+1$ (If $(k+p) = l = n+1$), then $B = \Lambda$ and $X_2 = \Lambda$.*

The case $j = k = l$ is not considered for parallel column contextual insertion operation.

Similarly, we can define *parallel row contextual insertion operation* by inserting row context I_r in between two sub-arrays A and B with the help of row operation $\ominus$, set of row array contexts R and row insertion partial function φ_r^i. We have $X \Rightarrow^{row_i} Y$ if $X = X_1 \ominus A \ominus B \ominus X_2$ and $Y = X_1 \ominus A \ominus I_r \ominus B \ominus X_2$.

Example 3. Consider the following array

$$Z = \begin{matrix} a & a & b & b \\ a & a & b & b \\ b & b & a & a \\ b & b & a & a \end{matrix}$$ along with the sets of column and row contexts (to be inserted) represented by $C = \left\{ \begin{matrix} a\,b \\ b\,a \end{matrix}, \begin{matrix} a\,b \\ a\,b \end{matrix}, \begin{matrix} b\,a \\ b\,a \end{matrix}, \begin{matrix} b\,b \\ a\,a \end{matrix} \right\}$ and $R = \left\{ \begin{matrix} a\,b \\ b\,a \end{matrix}, \begin{matrix} a\,a \\ b\,b \end{matrix} \right\}$ respectively.

We now define column and row insertion rules (using partial functions φ_c^i and φ_r^i) as follows:

$$\text{IR1} = \left\{ \varphi_c^i \begin{bmatrix} a & b \\ a, & b \end{bmatrix} = \begin{bmatrix} a\,b \\ a\,b \end{bmatrix}, \varphi_c^i \begin{bmatrix} a & b \\ b, & a \end{bmatrix} = \begin{bmatrix} a\,b \\ b\,a \end{bmatrix}, \varphi_c^i \begin{bmatrix} b & a \\ b, & a \end{bmatrix} = \begin{bmatrix} b\,a \\ b\,a \end{bmatrix} \right\}.$$

$$\text{IR2} = \left\{ \varphi_r^i \begin{bmatrix} a\,b, & b\,a \end{bmatrix} = \begin{bmatrix} a\,b \\ b\,a \end{bmatrix}, \varphi_r^i \begin{bmatrix} a\,a, & b\,b \end{bmatrix} = \begin{bmatrix} a\,a \\ b\,b \end{bmatrix} \right\}.$$

Now using column contexts and column insertion rules, a sample horizontal growth of the picture Z is given below:

$$Z = \begin{matrix} a\,a\,b\,b \\ a\,a\,b\,b \\ b\,b\,a\,a \\ b\,b\,a\,a \end{matrix} \Rightarrow_{IR1}^{col_i} \begin{matrix} a\,a\,a\,b\,b\,b \\ a\,a\,a\,b\,b\,b \\ b\,b\,b\,a\,a\,a \\ b\,b\,b\,a\,a\,a \end{matrix} = \begin{matrix} a\,a\,a\,b\,b\,b \\ a\,a\,a\,b\,b\,b \\ b\,b\,b\,a\,a\,a \\ b\,b\,b\,a\,a\,a \end{matrix},$$ where col_i stands for column insertion and IRk stands for the k-th insertion rule set from which the corresponding insertion rule (rules) is (are) taken.

Now we apply row insertion rules for the vertical growth of the previous picture as follows:

$$\begin{matrix} a\,a\,a\,b\,b\,b \\ a\,a\,a\,b\,b\,b \\ b\,b\,b\,a\,a\,a \\ b\,b\,b\,a\,a\,a \end{matrix} \Rightarrow_{IR2}^{row_i} \begin{matrix} a\,a\,a\,b\,b\,b \\ a\,a\,a\,b\,b\,b \\ a\,a\,a\,b\,b\,b \\ b\,b\,b\,a\,a\,a \\ b\,b\,b\,a\,a\,a \\ b\,b\,b\,a\,a\,a \end{matrix} = \begin{matrix} a\,a\,a\,b\,b\,b \\ a\,a\,a\,b\,b\,b \\ a\,a\,a\,b\,b\,b \\ b\,b\,b\,a\,a\,a \\ b\,b\,b\,a\,a\,a \\ b\,b\,b\,a\,a\,a \end{matrix}.$$

Definition 5. *The* parallel column contextual deletion operation *is defined as follows: Let V be an alphabet, C be a finite subset of V^{**} whose elements are the column array contexts and $\varphi_c^d : V^{**} \times V^{**} \to 2^C$ be a choice mapping. We define $\varphi_c^d : V^{**} \times V^{**} \to 2^{V^{**}}$ such that, for arrays* $A = \begin{bmatrix} a_{1j} & \cdots & a_{1(k-1)} \\ \vdots & \ddots & \vdots \\ a_{mj} & \cdots & a_{m(k-1)} \end{bmatrix}$,

$B = \begin{bmatrix} a_{1(k-p)} & \cdots & a_{1(l-1)} \\ \vdots & \ddots & \vdots \\ a_{m(k-p)} & \cdots & a_{m(l-1)} \end{bmatrix}$, $j < k < l, a_{ij} \in V$, $D_c \in \varphi_c^d(A, B), D_c = \begin{bmatrix} u_1 \\ u_2 \\ \vdots \\ u_m \end{bmatrix}$, *if*

$$c_i = \begin{bmatrix} u_i \\ u_{i+1} \end{bmatrix} \in \varphi_c^d \left(\begin{matrix} a_{ij} & \cdots & a_{i(k-1)} \\ a_{(i+1)j} & \cdots & a_{(i+1)(k-1)} \end{matrix}, \begin{matrix} a_{i(k+p)} & \cdots & a_{i(l-1)} \\ a_{(i+1)(k+p)} & \cdots & a_{(i+1)(l-1)} \end{matrix} \right),\ c_i \in$$

$C, 1 \le i \le m-1$, *not all need to be distinct.*

Given an array $X = [a_{ij}]_{m\times n}, a_{ij} \in V$, $X = X_1 \oplus A \oplus D_c \oplus B \oplus X_2$,

$$X_1 = \begin{bmatrix} a_{11} & \cdots & a_{1(j-1)} \\ \vdots & \ddots & \vdots \\ a_{m1} & \cdots & a_{m(j-1)} \end{bmatrix}, A = \begin{bmatrix} a_{1j} & \cdots & a_{1(k-1)} \\ \vdots & \ddots & \vdots \\ a_{mj} & \cdots & a_{m(k-1)} \end{bmatrix},$$

$$B = \begin{bmatrix} a_{1k} & \cdots & a_{1(l-1)} \\ \vdots & \ddots & \vdots \\ a_{mk} & \cdots & a_{m(l-1)} \end{bmatrix}, X_2 = \begin{bmatrix} a_{1l} & \cdots & a_{1n} \\ \vdots & \ddots & \vdots \\ a_{ml} & \cdots & a_{mn} \end{bmatrix}, 1 \le j \le k < l \le n+1 \text{ (or)}$$

$1 \le j < k \le l \le n+1$, *we write* $X \Rightarrow^{col_d} Y$ *if* $Y = X_1 \oplus A \oplus B \oplus X_2$, *such that* $D_c \in \varphi_c^d(A, B)$. D_c *is called the column context to be deleted. We say that* Y *is obtained from* X *by parallel column contextual insertion (deletion) operation. The following 4 special cases for* $X = X_1 \oplus A \oplus B \oplus X_2$ *are also considered,*

1. *For* $j = 1$ *we have* $X_1 = \Lambda$.
2. *For* $j = k$, *we have* $A = \Lambda$. *If* $j = k = 1$, *then* $X_1 = \Lambda$ *and* $A = \Lambda$.
3. *For* $k = l$ *(for* $k + p = l$*), we have* $B = \Lambda$.
4. *For* $l = n+1$, *we have* $X_2 = \Lambda$. *If* $k = l = n+1$ *(If* $(k+p) = l = n+1$*), then* $B = \Lambda$ *and* $X_2 = \Lambda$.

The case $j = k = l$ *is not considered for parallel column contextual deletion operation.*

Similarly, we can define *parallel row contextual deletion operation* by deleting row context D_r in between two sub-arrays A and B with the help of row operation $\ominus$ and set of row array contexts R. We have $X \Rightarrow^{row_d} Y$ if $X = X_1 \ominus A \ominus D_r \ominus B \ominus X_2$ and $Y = X_1 \ominus A \ominus B \ominus X_2$.

Example 4. Consider the following picture $M = \begin{matrix} a & a & Y & b & b \\ a & a & Y & b & b \\ X & X & Y & X & X \\ b & b & Y & a & a \\ b & b & Y & a & a \end{matrix}$ along with the column and row contexts (to be deleted) represented by $C = \left\{ \begin{matrix} Y \\ Y \end{matrix} \right\}$ and $R = \left\{ X \; X \right\}$ respectively. We now define column and row deletion rules (using partial functions φ_c^d and φ_r^d) as follows:

$$DR1 = \left\{ \varphi_c^d \begin{bmatrix} a & b \\ a & b \end{bmatrix} = \begin{bmatrix} Y \\ Y \end{bmatrix}, \varphi_c^d \begin{bmatrix} a & b \\ X & X \end{bmatrix} = \begin{bmatrix} Y \\ Y \end{bmatrix}, \varphi_c^d \begin{bmatrix} X & X \\ b & a \end{bmatrix} = \begin{bmatrix} Y \\ Y \end{bmatrix} \right\}.$$

$$DR2 = \left\{ \varphi_r^d \left[a\,a,\; b\,b \right] = \left[X\,X \right],\; \varphi_r^d \left[a\,b,\; b\,a \right] = \left[X\,X \right], \right.$$

$$\left. \varphi_r^d \left[b\,b,\; a\,a \right] = \left[X\,X \right] \right\}.$$

Now using column contexts and column deletion rules, to get rid of the symbols Y from the picture M is given as follows:

$$M = \begin{array}{ccccc} a & a & Y & b & b \\ a & a & Y & b & b \\ X & X & Y & X & X \\ b & b & Y & a & a \\ b & b & Y & a & a \end{array} \Rightarrow^{col_d}_{DR1} \begin{array}{cccc} a & a & b & b \\ a & a & b & b \\ X & X & X & X \\ b & b & a & a \\ b & b & a & a \end{array} = \begin{array}{cccc} a & a & b & b \\ a & a & b & b \\ X & X & X & X \\ b & b & a & a \\ b & b & a & a \end{array},$$ where col_d stands for column deletion and DRk stands for the k-th deletion rule set from which the corresponding deletion rule (rules) is (are) taken.

Now using row context and row deletion rules, to get rid of the symbols X from the previous picture as follows:

$$\begin{array}{cccc} a & a & b & b \\ a & a & b & b \\ X & X & X & X \\ b & b & a & a \\ b & b & a & a \end{array} \Rightarrow^{row_d}_{DR2} \begin{array}{cccc} a & a & b & b \\ a & a & b & b \\ b & b & a & a \\ b & b & a & a \end{array}.$$

Definition 6. *A* parallel contextual array insertion deletion grammar *is defined as* $G = (V, T, M, C, R, \varphi_c^i, \varphi_r^i, \varphi_c^d, \varphi_r^d)$, *where* V *is an alphabet,* $T \subseteq V$ *is a terminal alphabet,* M *is a finite subset of* V^{**} *called the base of* G, C *is a finite subset of* V^{**} *called column array contexts,* R *is a finite subset of* V^{**} *called row array contexts,* $\varphi_c^i : V^{**} \times V^{**} \to 2^C$, $\varphi_r^i : V^{**} \times V^{**} \to 2^R$, $\varphi_c^d : V^{**} \times V^{**} \to 2^C$, $\varphi_r^d : V^{**} \times V^{**} \to 2^R$, *are the choice mappings which perform the parallel column contextual insertion, parallel row contextual insertion, parallel column contextual deletion and parallel row contextual deletion operations, respectively.*

The one-step insertion derivation with respect to G *is a binary relation* $\Rightarrow_i$ *on* V^{**} *and is defined by* $X \Rightarrow_i Y$, *where* $X, Y \in V^{**}$ *if and only if* $X = X_1 \oplus A \oplus B \oplus X_2$, $Y = X_1 \oplus A \oplus I_c \oplus B \oplus X_2$ *or* $X = X_3 \ominus A \ominus B \ominus X_4$, $Y = X_3 \ominus A \ominus I_r \ominus B \ominus X_4$ *for some* $X_1, X_2, X_3, X_4 \in V^{**}$ *and* I_c, I_r *are inserted column and row contexts obtained by the parallel column or row contextual insertion operations according to the choice mappings.*

The one-step deletion derivation with respect to G *is a binary relation* $\Rightarrow_d$ *on* V^{**} *and is defined by* $X \Rightarrow_d Y$, *where* X, $Y \in V^{**}$ *if and only if* $X = X_1 \oplus A \oplus D_c \oplus B \oplus X_2$, $Y = X_1 \oplus A \oplus B \oplus X_2$ *or* $X = X_3 \ominus A \ominus D_r \ominus B \ominus X_4$, $Y = X_3 \ominus A \ominus B \ominus X_4$ *for some* $X_1, X_2, X_3, X_4 \in V^{**}$ *and* D_c, D_r *are column and row contexts to be deleted with respect to the parallel column or row contextual deletion operations according to the choice mappings.*

The direct derivation with respect to G *is a binary relation* $\Rightarrow_{i,d}$ *on* V^{**} *which is either* $\Rightarrow_i$ *or* $\Rightarrow_d$. *The reflexive transitive closure of* $\Rightarrow_{i,d}$ *is denoted by* $\Rightarrow_{i,d}^*$.

Definition 7. *Let* $G = (V, T, M, C, R, \varphi_c^i, \varphi_r^d, \varphi_c^d, \varphi_r^d)$ *be a parallel contextual array insertion deletion grammar (PCAIDG). The language generated by* G *is defined by,*

$$L(G) = \{Y \in T^{**} | \exists X \in M \text{ with } X \Rightarrow_{i,d}^* Y\}.$$

The family of all array languages generated by PCAIDGs is denoted by $\mathscr{L}(PCAIDG)$. We give two PCAIDGs to generate the language L_1 given in Example 1 and the language L_4 which is not generated by any TCSMG (see Example 6 below).

Example 5. Let $G_3 = \left(V, T, M, C, R, \varphi_c^i, \varphi_r^i, \varphi_c^d, \varphi_r^d\right)$, where

- $V = \{x, \bullet\}$,
- $T = \{x, \bullet\}$,
- $M = \left\{\begin{matrix} x \bullet x \\ x\, x\, x \\ x \bullet x \end{matrix}\right\}$,
- $C = \left\{\begin{matrix} \bullet \\ x \end{matrix}, \begin{matrix} \bullet \\ \bullet \end{matrix}, \begin{matrix} x \\ \bullet \end{matrix}\right\}$,
- $R = \left\{\bullet\, x\,, \bullet \bullet\,, x \bullet\right\}$,
- Sets of column and row insertion rules are given by

$$\text{IR1} = \left\{\varphi_c^i\begin{bmatrix} x & \bullet \\ x & x \end{bmatrix} = \begin{bmatrix} \bullet \\ x \end{bmatrix}, \varphi_c^i\begin{bmatrix} x & \bullet \\ x & \bullet \end{bmatrix} = \begin{bmatrix} \bullet \\ \bullet \end{bmatrix}, \varphi_c^i\begin{bmatrix} x & x \\ x & \bullet \end{bmatrix} = \begin{bmatrix} x \\ \bullet \end{bmatrix}\right\}.$$

$$\text{IR2} = \left\{\varphi_r^i\left[x \bullet\,,\ x\, x\right] = \left[x \bullet\right], \varphi_r^i\left[\bullet \bullet\,,\ x\, x\right] = \left[\bullet \bullet\right],\right.$$

$$\varphi_r^i\left[\bullet\, x\,,\ x\, x\right] = \left[\bullet\, x\right], \varphi_r^i\left[x\, x\,,\ x \bullet\right] = \left[x \bullet\right],$$

$$\left.\varphi_r^i\left[x\, x\,,\ \bullet \bullet\right] = \left[\bullet \bullet\right], \varphi_r^i\left[x\, x\,,\ \bullet\, x\right] = \left[\bullet\, x\right]\right\}.$$

$$\text{Here } L(G_3) = L_1 = \left\{\begin{matrix} x \bullet x \\ x\, x\, x \\ x \bullet x \end{matrix}, \begin{matrix} x \bullet x \\ x \bullet x \\ x\, x\, x \\ x \bullet x \end{matrix}, \begin{matrix} x \bullet x \\ x\, x\, x \\ x \bullet x \\ x \bullet x \end{matrix}, \begin{matrix} x \bullet \bullet x \\ x\, x\, x\, x \\ x \bullet \bullet x \end{matrix}, \begin{matrix} x \bullet x \\ x \bullet x \\ x\, x\, x \\ x \bullet x \\ x \bullet x \end{matrix}, \begin{matrix} x \bullet x \\ x\, x\, x \\ x \bullet x \\ x \bullet x \\ x \bullet x \end{matrix}, \cdots\right\}.$$

A sample derivation of a picture of size 7×5 in L_1 is given below: Here $P \Rightarrow_{IRk}^{\alpha_i} Q$ stands for $P \Rightarrow_i Q$, where $\alpha \in \{row, col\}$, the subscript i denotes insertion operation and IRk is the k-th insertion rules set. Similarly, $P \Rightarrow_{DRk}^{\beta_d} Q$ stands for $P \Rightarrow_d Q$, where $\beta \in \{row, col\}$, the subscript d denotes deletion operation and DRk is the k-th deletion rule set.

- is the (row or column) context inserted

$$\begin{array}{ccc} x & \bullet & x \\ x & x & x \\ x & \bullet & x \end{array} \Rightarrow_{IR1}^{col_i} \begin{array}{cccc} x & \bullet & \bullet & x \\ x & x & x & x \\ x & \bullet & \bullet & x \end{array} \Rightarrow_{IR1}^{col_i} \begin{array}{ccccc} x & \bullet & \bullet & \bullet & x \\ x & x & x & x & x \\ x & \bullet & \bullet & \bullet & x \end{array} \Rightarrow_{IR2}^{row_i}$$

$$\begin{array}{ccccc} x & \bullet & \bullet & \bullet & x \\ x & \bullet & \bullet & \bullet & x \\ x & x & x & x & x \\ x & \bullet & \bullet & \bullet & x \end{array} \Rightarrow_{IR2}^{row_i} \begin{array}{ccccc} x & \bullet & \bullet & \bullet & x \\ x & \bullet & \bullet & \bullet & x \\ x & \bullet & \bullet & \bullet & x \\ x & x & x & x & x \\ x & \bullet & \bullet & \bullet & x \end{array} \Rightarrow_{IR2}^{row_i} \begin{array}{ccccc} x & \bullet & \bullet & \bullet & x \\ x & \bullet & \bullet & \bullet & x \\ x & \bullet & \bullet & \bullet & x \\ x & x & x & x & x \\ x & \bullet & \bullet & \bullet & x \\ x & \bullet & \bullet & \bullet & x \end{array} \Rightarrow_{IR2}^{row_i} \begin{array}{ccccc} x & \bullet & \bullet & \bullet & x \\ x & \bullet & \bullet & \bullet & x \\ x & \bullet & \bullet & \bullet & x \\ x & x & x & x & x \\ x & \bullet & \bullet & \bullet & x \\ x & \bullet & \bullet & \bullet & x \\ x & \bullet & \bullet & \bullet & x \end{array} = \begin{array}{ccccc} x & \bullet & \bullet & \bullet & x \\ x & \bullet & \bullet & \bullet & x \\ x & \bullet & \bullet & \bullet & x \\ x & x & x & x & x \\ x & \bullet & \bullet & \bullet & x \\ x & \bullet & \bullet & \bullet & x \\ x & \bullet & \bullet & \bullet & x \end{array}.$$

Example 6. The language L_4 of H tokens of same proportion but of different sizes given by $\left\{ \begin{array}{l} \left(x \ (\bullet)^{2n-1} \ x \ \right)_{2n-2} \\ \left(x \ (x)^{2n-1} \ x \ \right) \\ \left(x \ (\bullet)^{2n-1} \ x \ \right)_{2n-2} \end{array} \middle| \ n \geq 2 \right\} \cup \left\{ \begin{array}{ccc} x & \bullet & x \\ x & x & x \\ x & \bullet & x \end{array} \right\}$ can neither be generated by a TCSMG [30] nor a CSMG [29]. We give a PCAIDG G_4 to generate the language L_4.

$$G_4 = \left(V, T, M, C, R, \varphi_c^i, \varphi_r^i, \varphi_c^d, \varphi_r^d \right), \text{ where}$$

- $V = \{x, Y, \bullet\}$;
- $T = \{x, \bullet\}$;
- $M = \left\{ \begin{array}{ccc} x & \bullet & x \\ x & x & x \\ x & \bullet & x \end{array} \right\}$;
- $C = \left\{ \begin{array}{cc} Y & \bullet \\ Y & x \end{array}, \begin{array}{cc} Y & \bullet \\ Y & \bullet \end{array}, \begin{array}{cc} Y & x \\ Y & \bullet \end{array}, \begin{array}{cc} \bullet & Y \\ x & Y \end{array}, \begin{array}{cc} \bullet & Y \\ \bullet & Y \end{array}, \begin{array}{cc} x & Y \\ \bullet & Y \end{array} \right\}$;
- $R = \left\{ \begin{array}{cc} x & Y \\ Y & Y \end{array}, \begin{array}{cc} Y & \bullet \\ Y & Y \end{array}, \begin{array}{cc} \bullet & \bullet \\ Y & Y \end{array}, \begin{array}{cc} \bullet & Y \\ Y & Y \end{array}, \begin{array}{cc} Y & x \\ Y & Y \end{array}, \begin{array}{cc} Y & Y \\ x & Y \end{array}, \begin{array}{cc} Y & Y \\ Y & \bullet \end{array}, \begin{array}{cc} Y & Y \\ \bullet & \bullet \end{array}, \begin{array}{cc} Y & Y \\ \bullet & Y \end{array}, \begin{array}{cc} Y & Y \\ Y & x \end{array} \right\}$;
- Sets of column and row insertion rules are given by

$$\begin{aligned} \text{IR1} = \Bigg\{ & \varphi_c^i \begin{bmatrix} x & \bullet \\ x \ , & x \end{bmatrix} = \begin{bmatrix} Y & \bullet \\ Y & x \end{bmatrix}, \ \varphi_c^i \begin{bmatrix} x & \bullet \\ x \ , & \bullet \end{bmatrix} = \begin{bmatrix} Y & \bullet \\ Y & \bullet \end{bmatrix}, \\ & \varphi_c^i \begin{bmatrix} x & x \\ x \ , & \bullet \end{bmatrix} = \begin{bmatrix} Y & x \\ Y & \bullet \end{bmatrix}, \ \varphi_c^i \begin{bmatrix} \bullet & x \\ x \ , & x \end{bmatrix} = \begin{bmatrix} \bullet & Y \\ x & Y \end{bmatrix}, \ \varphi_c^i \begin{bmatrix} \bullet & \bullet \\ x \ , & x \end{bmatrix} = \begin{bmatrix} \bullet & Y \\ \bullet & Y \end{bmatrix}, \\ & \varphi_c^i \begin{bmatrix} x & \bullet \\ x \ , & x \end{bmatrix} = \begin{bmatrix} x & Y \\ \bullet & Y \end{bmatrix} \Bigg\}. \end{aligned}$$

$$\text{IR2} = \Bigg\{ \varphi_r^i \begin{bmatrix} x \, Y \ , \ x \, Y \end{bmatrix} = \begin{bmatrix} x & Y \\ Y & Y \end{bmatrix}, \ \varphi_r^i \begin{bmatrix} Y \bullet , \ Y \, x \end{bmatrix} = \begin{bmatrix} Y & \bullet \\ Y & Y \end{bmatrix},$$

$$\varphi_r^i\left[\bullet\,\bullet\ ,\ x\,x\right]=\begin{bmatrix}\bullet & \bullet\\ Y & Y\end{bmatrix},\ \varphi_r^i\left[\bullet\,Y\ ,\ x\,Y\right]=\begin{bmatrix}\bullet & Y\\ Y & Y\end{bmatrix},$$

$$\varphi_r^i\left[Y\,x\ ,\ Y\,x\right]=\begin{bmatrix}Y & x\\ Y & Y\end{bmatrix},\ \varphi_r^i\left[x\,Y\ ,\ x\,Y\right]=\begin{bmatrix}Y & Y\\ x & Y\end{bmatrix},$$

$$\varphi_r^i\left[Y\,x\ ,\ Y\,\bullet\right]=\begin{bmatrix}Y & Y\\ Y & \bullet\end{bmatrix},\ \varphi_r^i\left[x\,x\ ,\ \bullet\,\bullet\right]=\begin{bmatrix}Y & Y\\ \bullet & \bullet\end{bmatrix},$$

$$\left.\varphi_r^i\left[x\,Y\ ,\ \bullet\,Y\right]=\begin{bmatrix}Y & Y\\ \bullet & Y\end{bmatrix},\ \varphi_r^i\left[Y\,x\ ,\ Y\,x\right]=\begin{bmatrix}Y & Y\\ Y & x\end{bmatrix}\right\}.$$

$$\mathrm{DR1}=\left\{\varphi_c^d\left[\begin{matrix}x\\x\end{matrix}\ ,\ \begin{matrix}\bullet\\\bullet\end{matrix}\right]=\begin{bmatrix}Y\\Y\end{bmatrix},\ \varphi_c^d\left[\begin{matrix}x\\Y\end{matrix}\ ,\ \begin{matrix}\bullet\\Y\end{matrix}\right]=\begin{bmatrix}Y\\Y\end{bmatrix},\right.$$

$$\varphi_c^d\left[\begin{matrix}Y\\x\end{matrix}\ ,\ \begin{matrix}Y\\x\end{matrix}\right]=\begin{bmatrix}Y\\Y\end{bmatrix},\ \varphi_c^d\left[\begin{matrix}x\\Y\end{matrix}\ ,\ \begin{matrix}x\\Y\end{matrix}\right]=\begin{bmatrix}Y\\Y\end{bmatrix},\ \varphi_c^d\left[\begin{matrix}Y\\x\end{matrix}\ ,\ \begin{matrix}Y\\\bullet\end{matrix}\right]=\begin{bmatrix}Y\\Y\end{bmatrix},$$

$$\varphi_c^d\left[\begin{matrix}\bullet\\\bullet\end{matrix}\ ,\ \begin{matrix}x\\x\end{matrix}\right]=\begin{bmatrix}Y\\Y\end{bmatrix},\ \varphi_c^d\left[\begin{matrix}\bullet\\Y\end{matrix}\ ,\ \begin{matrix}x\\Y\end{matrix}\right]=\begin{bmatrix}Y\\Y\end{bmatrix},\ \varphi_c^d\left[\begin{matrix}Y\\x\end{matrix}\ ,\ \begin{matrix}Y\\x\end{matrix}\right]=\begin{bmatrix}Y\\Y\end{bmatrix},$$

$$\left.\varphi_c^d\left[\begin{matrix}x\\Y\end{matrix}\ ,\ \begin{matrix}x\\Y\end{matrix}\right]=\begin{bmatrix}Y\\Y\end{bmatrix},\ \varphi_c^d\left[\begin{matrix}Y\\\bullet\end{matrix}\ ,\ \begin{matrix}Y\\x\end{matrix}\right]=\begin{bmatrix}Y\\Y\end{bmatrix}\right\}.$$

$$\mathrm{DR2}=\left\{\varphi_r^d\left[x\,\bullet\ ,\ x\,x\right]=\left[Y\,Y\right],\ \varphi_r^d\left[\bullet\,\bullet\ ,\ x\,x\right]=\left[Y\,Y\right],\right.$$

$$\varphi_r^d\left[\bullet\,x\ ,\ x\,x\right]=\left[Y\,Y\right],\ \varphi_r^d\left[x\,x\ ,\ x\,\bullet\right]=\left[Y\,Y\right],$$

$$\left.\varphi_r^d\left[x\,x\ ,\ \bullet\,\bullet\right]=\left[Y\,Y\right],\ \varphi_r^d\left[x\,x\ ,\ \bullet\,x\right]=\left[Y\,Y\right]\right\}.$$

Starting from the element of the axiom set M, a sample derivation of a picture in L_4 is given below:

- is the (row or column) context inserted

- is the (row or column) context to be deleted

$$\begin{matrix}x & \bullet & x\\ x & x & x\\ x & \bullet & x\end{matrix}\ \Rightarrow_{IR1}^{col_i}\ \begin{matrix}x & Y & \bullet & \bullet & x\\ x & Y & x & x & x\\ x & Y & \bullet & \bullet & x\end{matrix}\ =\ \begin{matrix}x & Y & \bullet & \bullet & x\\ x & Y & x & x & x\\ x & Y & \bullet & \bullet & x\end{matrix}\ \Rightarrow_{IR1}^{col_i}$$

$$\begin{matrix}x & Y & \bullet & \bullet & \bullet & Y & x\\ x & Y & x & x & x & Y & x\\ x & Y & \bullet & \bullet & \bullet & Y & x\end{matrix}\ =\ \begin{matrix}x & Y & \bullet & \bullet & \bullet & Y & x\\ x & Y & x & x & x & Y & x\\ x & Y & \bullet & \bullet & \bullet & Y & x\end{matrix}\ \Rightarrow_{IR2}^{row_i}\ \begin{matrix}x & Y & \bullet & \bullet & \bullet & Y & x\\ x & Y & \bullet & \bullet & \bullet & Y & x\\ Y & Y & Y & Y & Y & Y & Y\\ x & Y & x & x & x & Y & x\\ x & Y & \bullet & \bullet & \bullet & Y & x\end{matrix}\ =$$

$$\begin{matrix} x & Y & \bullet & \bullet & \bullet & Y & x \\ x & Y & \bullet & \bullet & \bullet & Y & x \\ Y & Y & Y & Y & Y & Y & Y \\ x & Y & x & x & x & Y & x \\ x & Y & \bullet & \bullet & \bullet & Y & x \end{matrix} \Rightarrow^{row_i}_{IR2} \begin{matrix} x & Y & \bullet & \bullet & \bullet & Y & x \\ x & Y & \bullet & \bullet & \bullet & Y & x \\ Y & Y & Y & Y & Y & Y & Y \\ x & Y & x & x & x & Y & x \\ Y & Y & Y & Y & Y & Y & Y \\ x & Y & \bullet & \bullet & \bullet & Y & x \\ x & Y & \bullet & \bullet & \bullet & Y & x \end{matrix} = \begin{matrix} x & Y & \bullet & \bullet & \bullet & Y & x \\ x & Y & \bullet & \bullet & \bullet & Y & x \\ Y & Y & Y & Y & Y & Y & Y \\ x & Y & x & x & x & Y & x \\ Y & Y & Y & Y & Y & Y & Y \\ x & Y & \bullet & \bullet & \bullet & Y & x \\ x & Y & \bullet & \bullet & \bullet & Y & x \end{matrix} \Rightarrow^{col_d}_{DR1}$$

$$\begin{matrix} x & \bullet & \bullet & \bullet & Y & x \\ x & \bullet & \bullet & \bullet & Y & x \\ Y & Y & Y & Y & Y & Y \\ x & x & x & x & Y & x \\ Y & Y & Y & Y & Y & Y \\ x & \bullet & \bullet & \bullet & Y & x \\ x & \bullet & \bullet & \bullet & Y & x \end{matrix} \Rightarrow^{col_d}_{DR1} \begin{matrix} x & \bullet & \bullet & \bullet & x \\ x & \bullet & \bullet & \bullet & x \\ Y & Y & Y & Y & Y \\ x & x & x & x & x \\ Y & Y & Y & Y & Y \\ x & \bullet & \bullet & \bullet & x \\ x & \bullet & \bullet & \bullet & x \end{matrix} \Rightarrow^{row_d}_{DR2} \begin{matrix} x & \bullet & \bullet & \bullet & x \\ x & \bullet & \bullet & \bullet & x \\ x & x & x & x & x \\ Y & Y & Y & Y & Y \\ x & \bullet & \bullet & \bullet & x \\ x & \bullet & \bullet & \bullet & x \end{matrix} \Rightarrow^{row_d}_{DR2}$$

$$\begin{matrix} x & \bullet & \bullet & \bullet & x \\ x & \bullet & \bullet & \bullet & x \\ x & x & x & x & x \\ x & \bullet & \bullet & \bullet & x \\ x & \bullet & \bullet & \bullet & x \end{matrix}.$$

Definition 8. *A* parallel contextual array insertion deletion P system with h membranes *($PCAIDPS_h$) is a construct,*

$$\prod = (V, T, \mu, C, R, (M_1, I_1, D_1), \ldots, (M_h, I_h, D_h), \varphi_c^i, \varphi_r^i, \varphi_c^d, \varphi_r^d, i_0)$$

where,

- *V is the finite nonempty set of symbols called alphabet;*
- *$T \subseteq V$ is the output alphabet;*
- *μ is the membrane structure with h membranes or regions;*
- *C is the finite subset of V^{**} called set of column array contexts;*
- *R is the finite subset of V^{**} called set of row array contexts;*
- *M_i is the finite set of arrays over V called as axioms associated with the region μ_i of μ;*
- *$\varphi_c^i : V^{**} \times V^{**} \rightarrow 2^C$ is the partial mapping performing parallel column contextual insertion operations;*
- *$\varphi_r^i : V^{**} \times V^{**} \rightarrow 2^R$ is the partial mapping performing parallel row contextual insertion operations;*
- *$\varphi_c^d : V^{**} \times V^{**} \rightarrow 2^C$ is the partial mapping performing parallel column contextual deletion operations;*
- *$\varphi_r^d : V^{**} \times V^{**} \rightarrow 2^R$ is the partial mapping performing parallel row contextual deletion operations;*

- $I_i = \emptyset$ *(or)* $\left\{ \left(\left\{ \varphi_c^i(A_i, B_i) = \begin{bmatrix} u_i \\ u_{i+1} \end{bmatrix} \middle| i = 1, 2, \ldots, m-1 \right\}, \alpha \right) \right\}$ *where* $A_i = \begin{bmatrix} a_{ij} & \cdots & a_{i(k-1)} \\ a_{(i+1)j} & \cdots & a_{(i+1)(k-1)} \end{bmatrix}, B_i = \begin{bmatrix} a_{ik} & \cdots & a_{i(l-1)} \\ a_{(i+1)k} & \cdots & a_{(i+1)(l-1)} \end{bmatrix}$, $1 \leq j \leq k < l \leq n+1$ *(or)* $1 \leq j < k \leq l \leq n+1$, $\alpha \in \{here, out, in_t\}$, *$u_i$ and u_{i+1} are arrays of size $1 \times p$ with $p \geq 1$.*
  (or)
  $\left\{ \left(\left\{ \varphi_r^i(C_i, E_i) = [u_i \; u_{i+1}] \middle| i = 1, 2, \ldots, n-1 \right\}, \alpha \right) \right\}$ *where* $C_i = \begin{bmatrix} a_{ji} & a_{j(i+1)} \\ \vdots & \vdots \\ a_{(k-1)i} & a_{(k-1)(i+1)} \end{bmatrix}, E_i = \begin{bmatrix} a_{ki} & a_{k(i+1)} \\ \vdots & \vdots \\ a_{(l-1)i} & a_{(l-1)(i+1)} \end{bmatrix}$, $1 \leq j \leq k < l \leq m+1$ *(or)* $1 \leq j < k \leq l \leq m+1$, $\alpha \in \{here, out, in_t\}$, *where 'here' stands for the current membrane where the actions are being performed, 'out' stands for immediate outer membrane of the current membrane and 'in_t' stands for the specified membrane t to which the controls are to be transferred to. Here u_i and u_{i+1} are arrays of size $p \times 1$ with $p \geq 1$.*
- $D_i = \emptyset$ *(or)* $\left\{ \left(\left\{ \varphi_c^d(A_i, B_i) = \begin{bmatrix} u_i \\ u_{i+1} \end{bmatrix} \middle| i = 1, 2, \ldots, m-1 \right\}, \alpha \right) \right\}$ *where* $A_i = \begin{bmatrix} a_{ij} & \cdots & a_{i(k-1)} \\ a_{(i+1)j} & \cdots & a_{(i+1)(k-1)} \end{bmatrix}, B_i = \begin{bmatrix} a_{i(k+p)} & \cdots & a_{i(l-1)} \\ a_{(i+1)(k+p)} & \cdots & a_{(i+1)(l-1)} \end{bmatrix}$, $1 \leq j \leq k < l \leq n+1$, $\alpha \in \{here, out, in_t\}$, *$u_i$ and u_{i+1} are arrays of size $1 \times p$ with $p \geq 1$.*
  (or)
  $\left\{ \left(\left\{ \varphi_r^d(C_i, E_i) = [u_i \; u_{i+1}] \middle| i = 1, 2, \ldots, n-1 \right\}, \alpha \right) \right\}$ *where* $C_i = \begin{bmatrix} a_{ji} & a_{j(i+1)} \\ \vdots & \vdots \\ a_{(k-1)i} & a_{(k-1)(i+1)} \end{bmatrix}, E_i = \begin{bmatrix} a_{(k+p)i} & a_{(k+p)(i+1)} \\ \vdots & \vdots \\ a_{(l-1)i} & a_{(l-1)(i+1)} \end{bmatrix}$, $1 \leq j \leq k < l \leq m+1$, $\alpha \in \{here, out, in_t\}$, *$u_i$ and u_{i+1} are arrays of size $p \times 1$ with $p \geq 1$.*

- *i_0 is the output membrane.*

The array language generated by $\prod$ is denoted by $L(\prod)$ and the family of array languages generated by PCAIDPS with h membranes is denoted by $\mathscr{L}(PCAIDPS_h)$. We give two PCAIDPS$_2$ to generate the array language L_1 given in Example 1 and the language L_4 given in Example 6.

Example 7. $\prod_1 = \left(V, T, \mu, C, R, (M_1, I_1, D_1), (M_2, I_2, D_2), \varphi_c^i, \varphi_r^i, \varphi_c^d, \varphi_r^d, i_0 \right)$,

where

- $V = \{x, \bullet\}$;
- $T = \{x, \bullet\}$;
- $\mu = [_2[_1]_1]_2$;
- $C = \left\{ \begin{matrix} \bullet & \bullet \\ x & \bullet \end{matrix}, \begin{matrix} x \\ \bullet \end{matrix} \right\}$;

- $R = \left\{ \bullet\, x \;, \bullet\, \bullet \;, x\, \bullet \right\}$;
- $M_1 = \emptyset$; $M_2 = \left\{ \begin{matrix} x & \bullet & x \\ x & x & x \\ x & \bullet & x \end{matrix} \right\}$;
- $I_1 = \emptyset$;
- $I_2 = \{IR1, IR2, IR3, IR4\}$ with
  - IR1 $= \left\{ \varphi_c^i \left[\begin{matrix} x \\ x \end{matrix} \;,\; \begin{matrix} \bullet \\ x \end{matrix} \right] = \left\{ \begin{matrix} \bullet \\ x \end{matrix} \right\},\ \varphi_c^i \left[\begin{matrix} x \\ x \end{matrix} \;,\; \begin{matrix} \bullet \\ \bullet \end{matrix} \right] = \left\{ \begin{matrix} \bullet \\ \bullet \end{matrix} \right\},\ \varphi_c^i \left[\begin{matrix} x \\ x \end{matrix} \;,\; \begin{matrix} x \\ \bullet \end{matrix} \right] = \left\{ \begin{matrix} x \\ \bullet \end{matrix} \right\} \right\}, here \right\}$.
  - IR2 $= \left\{ \varphi_r^i \left[x\, \bullet \;,\; x\, x \right] = \left\{ x\, \bullet \right\},\ \varphi_r^i \left[\bullet\, \bullet \;,\; x\, x \right] = \left\{ \bullet\, \bullet \right\},\ \varphi_r^i \left[\bullet\, x \;,\; x\, x \right] = \left\{ \bullet\, x \right\} \right\}, here \right\}$.
  - IR3 $= \left\{ \varphi_r^i \left[x\, x \;,\; x\, \bullet \right] = \left\{ x\, \bullet \right\},\ \varphi_r^i \left[x\, x \;,\; \bullet\, \bullet \right] = \left\{ \bullet\, \bullet \right\},\ \varphi_r^i \left[x\, x \;,\; \bullet\, x \right] = \left\{ \bullet\, x \right\} \right\}, here \right\}$.
  - IR4 $= \left\{ \varphi_c^i \left[\begin{matrix} x \\ x \end{matrix} \;,\; \begin{matrix} \bullet \\ x \end{matrix} \right] = \left\{ \begin{matrix} \lambda \\ \lambda \end{matrix} \right\},\ \varphi_c^i \left[\begin{matrix} x \\ x \end{matrix} \;,\; \begin{matrix} \bullet \\ \bullet \end{matrix} \right] = \left\{ \begin{matrix} \lambda \\ \lambda \end{matrix} \right\},\ \varphi_c^i \left[\begin{matrix} x \\ x \end{matrix} \;,\; \begin{matrix} x \\ \bullet \end{matrix} \right] = \left\{ \begin{matrix} \lambda \\ \lambda \end{matrix} \right\} \right\}, out \right\}$.
- $D_1 = \emptyset$;
- $D_2 = \emptyset$;
- i_0 is the membrane 1.

A sample derivation of a picture of size 7×5 in L_1 is given as follows:

- is the (row or column) context inserted

$$\begin{matrix} x & \bullet & x \\ x & x & x \\ x & \bullet & x \end{matrix} \Rightarrow^{col_i}_{(2,IR1)} \begin{matrix} x & \bullet & \bullet & \bullet & x \\ x & x & x & x & x \\ x & \bullet & \bullet & \bullet & x \end{matrix} \Rightarrow^{row_i}_{(2,IR2)} \begin{matrix} x & \bullet & \bullet & \bullet & x \\ x & \bullet & \bullet & \bullet & x \\ x & x & x & x & x \\ x & \bullet & \bullet & \bullet & x \end{matrix} \Rightarrow^{row_i}_{(2,IR2)}$$

$$\begin{array}{ccccccc} & & & & x \bullet \bullet \bullet x & & x \bullet \bullet \bullet x \\ & & x \bullet \bullet \bullet x & & x \bullet \bullet \bullet x & & x \bullet \bullet \bullet x \\ x \bullet \bullet \bullet x & & x \bullet \bullet \bullet x & & x \bullet \bullet \bullet x & & x \bullet \bullet \bullet x \\ x \bullet \bullet \bullet x & \Rightarrow^{row_i}_{(2,IR2)} & x \bullet \bullet \bullet x & \Rightarrow^{row_i}_{(2,IR2)} & x\,x\,x\,x\,x & \Rightarrow^{col_i}_{(2,IR4)} & x\,x\,x\,x\,x. \\ x \bullet \bullet \bullet x & & x\,x\,x\,x\,x & & x \bullet \bullet \bullet x & & x \bullet \bullet \bullet x \\ x\,x\,x\,x\,x & & x \bullet \bullet \bullet x & & x \bullet \bullet \bullet x & & x \bullet \bullet \bullet x \\ x \bullet \bullet \bullet x & & x \bullet \bullet \bullet x & & x \bullet \bullet \bullet x & & x \bullet \bullet \bullet x \end{array}$$

Note 1. Here $P \Rightarrow^{\alpha_i}_{(n,IRk)} Q$ stands for $P \Rightarrow_i Q$, where $\alpha \in \{row, col\}$, subscript i denotes insertion operation , n is the nth membrane in which insertion operation is being performed and IRk is the k-th insertion rules set. Similarly, $P \Rightarrow^{\beta_d}_{(n,DRk)} Q$ stands for $P \Rightarrow_d Q$, where $\beta \in \{row, col\}$, subscript d denotes deletion operation, n is the nth membrane in which the deletion operation is being performed and DRk is the k-th deletion rules set.

Example 8. We give a PCAIDPS$_2$ $\prod_2$ to generate the array language

$$L_4 = \left\{ \begin{array}{c} \left(x\ (\bullet)^{2n-1}\ x\right)_{2n-2} \\ x\ \ (x)^{2n-1}\ \ x \\ \left(x\ (\bullet)^{2n-1}\ x\right)_{2n-2} \end{array} \middle|\ n \geq 2 \right\} \cup \left\{ \begin{array}{c} x \bullet x \\ x\,x\,x \\ x \bullet x \end{array} \right\}.$$

Let $\prod_2 = \left(V, T, \mu, C, R, (M_1, I_1, D_1), (M_2, I_2, D_2), \varphi^i_c, \varphi^i_r, \varphi^d_c, \varphi^d_r, i_0 \right)$, where

- $V = \{x, Y, \bullet\}$; $T = \{x, \bullet\}$; $\mu = [_1[_2]_2]_1$; $M_2 = \left\{ \begin{array}{c} x \bullet x \\ x\,x\,x \\ x \bullet x \end{array} \right\}$; $M_1 = \emptyset$;

- $C = \left\{ \begin{array}{c} Y\,\bullet \\ Y\,x \end{array}, \begin{array}{c} Y\,\bullet \\ Y\,\bullet \end{array}, \begin{array}{c} Y\,x \\ Y\,\bullet \end{array}, \begin{array}{c} \bullet\,Y \\ x\,Y \end{array}, \begin{array}{c} \bullet\,Y \\ \bullet\,Y \end{array}, \begin{array}{c} x\,Y \\ \bullet\,Y \end{array} \right\}$;

- $R = \left\{ \begin{array}{c} x\,Y \\ Y\,Y \end{array}, \begin{array}{c} Y\,\bullet \\ Y\,Y \end{array}, \begin{array}{c} \bullet\,\bullet \\ Y\,Y \end{array}, \begin{array}{c} \bullet\,Y \\ Y\,Y \end{array}, \begin{array}{c} Y\,x \\ Y\,Y \end{array}, \begin{array}{c} Y\,Y \\ x\,Y \end{array}, \begin{array}{c} Y\,Y \\ Y\,\bullet \end{array}, \begin{array}{c} Y\,Y \\ \bullet\,\bullet \end{array}, \begin{array}{c} Y\,Y \\ \bullet\,Y \end{array}, \begin{array}{c} Y\,Y \\ Y\,x \end{array} \right\}$;

Sets of column and row insertion rules of I_2 in membrane 2 are defined by

$$\text{IR1} = \left\{ \left\{ \varphi^i_c \begin{bmatrix} x & \bullet \\ x & x \end{bmatrix} = \begin{bmatrix} Y\,\bullet \\ Y\,x \end{bmatrix}, \varphi^i_c \begin{bmatrix} x & \bullet \\ x & \bullet \end{bmatrix} = \begin{bmatrix} Y\,\bullet \\ Y\,\bullet \end{bmatrix}, \right. \right.$$

$$\left. \left. \varphi^i_c \begin{bmatrix} x & x \\ x & \bullet \end{bmatrix} = \begin{bmatrix} Y\,x \\ Y\,\bullet \end{bmatrix} \right\}, here \right\}.$$

$$\text{IR2} = \left\{ \left\{ \varphi^i_c \begin{bmatrix} \bullet & x \\ x & x \end{bmatrix} = \begin{bmatrix} \bullet\,Y \\ x\,Y \end{bmatrix}, \varphi^i_c \begin{bmatrix} \bullet & x \\ \bullet & x \end{bmatrix} = \begin{bmatrix} \bullet\,Y \\ \bullet\,Y \end{bmatrix}, \right. \right.$$

$$\left. \left. \varphi^i_c \begin{bmatrix} x & x \\ \bullet & x \end{bmatrix} = \begin{bmatrix} x\,Y \\ \bullet\,Y \end{bmatrix} \right\}, here \right\}.$$

$$\text{IR3} = \Bigg\{\Bigg\{\varphi_r^i\begin{bmatrix} x\,Y \;,\; x\,Y \end{bmatrix} = \begin{bmatrix} x & Y \\ Y & Y \end{bmatrix},\ \varphi_r^i\begin{bmatrix} Y\,\bullet \;,\; Y\,x \end{bmatrix} = \begin{bmatrix} Y & \bullet \\ Y & Y \end{bmatrix},$$

$$\varphi_r^i\begin{bmatrix} \bullet\,\bullet \;,\; x\,x \end{bmatrix} = \begin{bmatrix} \bullet & \bullet \\ Y & Y \end{bmatrix},\ \varphi_r^i\begin{bmatrix} \bullet\,Y \;,\; x\,Y \end{bmatrix} = \begin{bmatrix} \bullet & Y \\ Y & Y \end{bmatrix},$$

$$\varphi_r^i\begin{bmatrix} Y\,x \;,\; Y\,x \end{bmatrix} = \begin{bmatrix} Y & x \\ Y & Y \end{bmatrix}\Bigg\}, here\Bigg\}.$$

$$\text{IR4} = \Bigg\{\Bigg\{\varphi_r^i\begin{bmatrix} x\,Y \;,\; x\,Y \end{bmatrix} = \begin{bmatrix} Y & Y \\ x & Y \end{bmatrix},\varphi_r^i\begin{bmatrix} Y\,x \;,\; Y\,\bullet \end{bmatrix} = \begin{bmatrix} Y & Y \\ Y & \bullet \end{bmatrix},$$

$$\varphi_r^i\begin{bmatrix} x\,x \;,\; \bullet\,\bullet \end{bmatrix} = \begin{bmatrix} Y & Y \\ \bullet & \bullet \end{bmatrix},\ \varphi_r^i\begin{bmatrix} x\,Y \;,\; \bullet\,Y \end{bmatrix} = \begin{bmatrix} Y & Y \\ \bullet & Y \end{bmatrix},$$

$$\varphi_r^i\begin{bmatrix} Y\,x \;,\; Y\,x \end{bmatrix} = \begin{bmatrix} Y & Y \\ Y & x \end{bmatrix}\Bigg\}, here\Bigg\}.$$

Sets of column and row deletion rules of D_2 in membrane 2 are defined by

$$\text{DR1} = \Bigg\{\Bigg\{\varphi_c^d\begin{bmatrix} x & & \bullet \\ x & , & \bullet \end{bmatrix} = \begin{bmatrix} Y \\ Y \end{bmatrix},\ \varphi_c^d\begin{bmatrix} x & & \bullet \\ Y & , & Y \end{bmatrix} = \begin{bmatrix} Y \\ Y \end{bmatrix},\ \varphi_c^d\begin{bmatrix} Y & & Y \\ x & , & x \end{bmatrix} = \begin{bmatrix} Y \\ Y \end{bmatrix},$$

$$\varphi_c^d\begin{bmatrix} x & & x \\ Y & , & Y \end{bmatrix} = \begin{bmatrix} Y \\ Y \end{bmatrix},\ \varphi_c^d\begin{bmatrix} Y & & Y \\ x & , & \bullet \end{bmatrix} = \begin{bmatrix} Y \\ Y \end{bmatrix}\Bigg\}, here\Bigg\}.$$

$$\text{DR2} = \Bigg\{\Bigg\{\varphi_c^d\begin{bmatrix} \bullet & & x \\ \bullet & , & x \end{bmatrix} = \begin{bmatrix} Y \\ Y \end{bmatrix},\ \varphi_c^d\begin{bmatrix} \bullet & & x \\ Y & , & Y \end{bmatrix} = \begin{bmatrix} Y \\ Y \end{bmatrix},\ \varphi_c^d\begin{bmatrix} Y & & Y \\ x & , & x \end{bmatrix} = \begin{bmatrix} Y \\ Y \end{bmatrix},$$

$$\varphi_c^d\begin{bmatrix} x & & x \\ Y & , & Y \end{bmatrix} = \begin{bmatrix} Y \\ Y \end{bmatrix},\ \varphi_c^d\begin{bmatrix} Y & & Y \\ \bullet & , & x \end{bmatrix} = \begin{bmatrix} Y \\ Y \end{bmatrix}\Bigg\}, here\Bigg\}.$$

$$\text{DR3} = \Bigg\{\Bigg\{\varphi_r^d\begin{bmatrix} x\,\bullet \;,\; x\,x \end{bmatrix} = \begin{bmatrix} Y\,Y \end{bmatrix},\ \varphi_r^d\begin{bmatrix} \bullet\,\bullet \;,\; x\,x \end{bmatrix} = \begin{bmatrix} Y\,Y \end{bmatrix},$$

$$\varphi_r^d\begin{bmatrix} \bullet\,x \;,\; x\,x \end{bmatrix} = \begin{bmatrix} Y\,Y \end{bmatrix}\Bigg\}, \alpha\Bigg\}, \text{ where } \alpha \in \{here, out\}.$$

$$\text{DR4} = \Bigg\{\varphi_r^d\begin{bmatrix} x\,x \;,\; x\,\bullet \end{bmatrix} = \begin{bmatrix} Y\,Y \end{bmatrix},\ \varphi_r^d\begin{bmatrix} x\,x \;,\; \bullet\,\bullet \end{bmatrix} = \begin{bmatrix} Y\,Y \end{bmatrix},$$

$$\varphi_r^d\begin{bmatrix} x\,x \;,\; \bullet\,x \end{bmatrix} = \begin{bmatrix} Y\,Y \end{bmatrix}\Bigg\}, \alpha\Bigg\}, \text{ where } \alpha \in \{here, out\}.$$

Set of deletion rules of I_1 in membrane M_1 is given by

$$\text{DR5} = \Bigg\{\Bigg\{\varphi_r^d\begin{bmatrix} x\,\bullet \;,\; x\,x \end{bmatrix} = \begin{bmatrix} Y\,Y \end{bmatrix},\ \varphi_r^d\begin{bmatrix} \bullet\,\bullet \;,\; x\,x \end{bmatrix} = \begin{bmatrix} Y\,Y \end{bmatrix},$$

$$\varphi_r^d\begin{bmatrix} \bullet\,x \;,\; x\,x \end{bmatrix} = \begin{bmatrix} Y\,Y \end{bmatrix},\varphi_r^d\begin{bmatrix} x\,x \;,\; x\,\bullet \end{bmatrix} = \begin{bmatrix} Y\,Y \end{bmatrix},$$

$$\varphi_r^d\begin{bmatrix} x\,x \;,\; \bullet\,\bullet \end{bmatrix} = \begin{bmatrix} Y\,Y \end{bmatrix},\ \varphi_r^d\begin{bmatrix} x\,x \;,\; \bullet\,x \end{bmatrix} = \begin{bmatrix} Y\,Y \end{bmatrix}\Bigg\}, here\Bigg\}.$$

Starting from the array in axiom set M_2, a sample computation of a picture of size 7×5 in L_4 is given below:

(dark grey) - is the (row or column) context inserted

(light grey) - is the (row or column) context to be deleted

$$\begin{array}{ccc} x & \bullet & x \\ x & x & x \\ x & \bullet & x \end{array} \Rightarrow^{col_i}_{(2,IR1)} \begin{array}{ccccc} x & Y & \bullet & \bullet & x \\ x & Y & x & x & x \\ x & Y & \bullet & \bullet & x \end{array} = \begin{array}{ccccc} x & Y & \bullet & \bullet & x \\ x & Y & x & x & x \\ x & Y & \bullet & \bullet & x \end{array} \Rightarrow^{col_i}_{(2,IR1)}$$

$$\begin{array}{ccccccc} x & Y & \bullet & \bullet & \bullet & Y & x \\ x & Y & x & x & x & Y & x \\ x & Y & \bullet & \bullet & \bullet & Y & x \end{array} = \begin{array}{ccccccc} x & Y & \bullet & \bullet & \bullet & Y & x \\ x & Y & x & x & x & Y & x \\ x & Y & \bullet & \bullet & \bullet & Y & x \end{array} \Rightarrow^{row_i}_{(2,IR1)} \begin{array}{ccccccc} x & Y & \bullet & \bullet & \bullet & Y & x \\ x & Y & \bullet & \bullet & \bullet & Y & x \\ Y & Y & Y & Y & Y & Y & Y \\ x & Y & x & x & x & Y & x \\ x & Y & \bullet & \bullet & \bullet & Y & x \end{array} =$$

$$\begin{array}{ccccccc} x & Y & \bullet & \bullet & \bullet & Y & x \\ x & Y & \bullet & \bullet & \bullet & Y & x \\ Y & Y & Y & Y & Y & Y & Y \\ x & Y & x & x & x & Y & x \\ x & Y & \bullet & \bullet & \bullet & Y & x \end{array} \Rightarrow^{row_i}_{(2,IR1)} \begin{array}{ccccccc} x & Y & \bullet & \bullet & \bullet & Y & x \\ x & Y & \bullet & \bullet & \bullet & Y & x \\ Y & Y & Y & Y & Y & Y & Y \\ x & Y & x & x & x & Y & x \\ Y & Y & Y & Y & Y & Y & Y \\ x & Y & \bullet & \bullet & \bullet & Y & x \\ x & Y & \bullet & \bullet & \bullet & Y & x \end{array} = \begin{array}{ccccccc} x & Y & \bullet & \bullet & \bullet & Y & x \\ x & Y & \bullet & \bullet & \bullet & Y & x \\ Y & Y & Y & Y & Y & Y & Y \\ x & Y & x & x & x & Y & x \\ Y & Y & Y & Y & Y & Y & Y \\ x & Y & \bullet & \bullet & \bullet & Y & x \\ x & Y & \bullet & \bullet & \bullet & Y & x \end{array} \Rightarrow^{col_d}_{(2,DR1)}$$

$$\begin{array}{cccccc} x & \bullet & \bullet & \bullet & Y & x \\ x & \bullet & \bullet & \bullet & Y & x \\ Y & Y & Y & Y & Y & Y \\ x & x & x & x & Y & x \\ Y & Y & Y & Y & Y & Y \\ x & \bullet & \bullet & \bullet & Y & x \\ x & \bullet & \bullet & \bullet & Y & x \end{array} \Rightarrow^{col_d}_{(2,DR1)} \begin{array}{ccccc} x & \bullet & \bullet & \bullet & x \\ x & \bullet & \bullet & \bullet & x \\ Y & Y & Y & Y & Y \\ x & x & x & x & x \\ Y & Y & Y & Y & Y \\ x & \bullet & \bullet & \bullet & x \\ x & \bullet & \bullet & \bullet & x \end{array} = \begin{array}{ccccc} x & \bullet & \bullet & \bullet & x \\ x & \bullet & \bullet & \bullet & x \\ Y & Y & Y & Y & Y \\ x & x & x & x & x \\ Y & Y & Y & Y & Y \\ x & \bullet & \bullet & \bullet & x \\ x & \bullet & \bullet & \bullet & x \end{array} \Rightarrow^{row_d}_{(2,DR1)}$$

$$\begin{array}{ccccc} x & \bullet & \bullet & \bullet & x \\ x & \bullet & \bullet & \bullet & x \\ x & x & x & x & x \\ Y & Y & Y & Y & Y \\ x & \bullet & \bullet & \bullet & x \\ x & \bullet & \bullet & \bullet & x \end{array} = \begin{array}{ccccc} x & \bullet & \bullet & \bullet & x \\ x & \bullet & \bullet & \bullet & x \\ x & x & x & x & x \\ Y & Y & Y & Y & Y \\ x & \bullet & \bullet & \bullet & x \\ x & \bullet & \bullet & \bullet & x \end{array} \Rightarrow^{row_d}_{(2,DR2)} \begin{array}{ccccc} x & \bullet & \bullet & \bullet & x \\ x & \bullet & \bullet & \bullet & x \\ x & x & x & x & x \\ x & \bullet & \bullet & \bullet & x \\ x & \bullet & \bullet & \bullet & x \end{array}.$$

4 Results

In this section, we compare the generative powers of PCAIDG and PCAIDPS$_h$ with TCSMG.

Theorem 1. $\mathscr{L}(TCSMG) \subsetneq \mathscr{L}(PCAIDG)$

The idea of the proof is as follows:
We show that for every TCSMG generating a language L we can construct a corresponding PCAIDG to generate the same language L. In the PCAIDG that we construct, we start with an axiom $\begin{matrix} \# & \# & \# \\ \# & S & \# \end{matrix}$ corresponding to the start symbol S in the TCSMG. For each horizontal rule in the TCSMG we define one insertion rule and one deletion rule in the PCAIDG which have the same effect as that of the rule considered. Similarly, for each vertical rule in an arbitrary table of TCSMG we accordingly define sets of insertion and deletion rules in PCAIDG having the same effect as the vertical rule considered. Finally we define deletion rules in PCAIDG to delete all the remaining #'s in the pictures generated.

Proof. We prove the theorem using the construction given below: Consider an arbitrary TCSMG $G_T = (G_H, G_V)$ where $G_H = (N_H, I, P_H, S)$ is a context-sensitive grammar with

- N_H, a finite set of horizontal non-terminals
- $I = \{S_1, \ldots, S_k\}$, a finite set of intermediates, $N_H \cap I = \emptyset$
- P_H, a finite set of context-sensitive horizontal rules
- $S \in N_H$ is the start symbol and
- $G_V = \left(\bigcup_{i=1}^{k} G_i, \mathscr{P}\right)$ where $G_i = (N_i, T', P_i, S_i)$, $i = 1, \ldots, k$ are right-linear grammars with
  - N_i, a finite set of vertical non-terminals; $N_i \cap N_j = \emptyset$ if $i \neq j$
  - T', a finite set of terminals
  - $P_i = P_{N_i} \cup P_{T_i}$
  - P_{N_i}, a finite set of right linear non-terminal production rules of the form $A \to aB$;
  - P_{T_i}, a finite set of right linear production rules of the form $A \to a$; $A, B \in N_i$, $a \in T \cup \{\lambda\}$
  - $S_i \in N_i$, the start symbol of G_i
  - $\mathscr{P} = \{t_1, t_2, \ldots t_n\}$, a finte set of tables.

We construct a PCAIDG $G = \left(V, T, M, C, R, \varphi_c^i, \varphi_r^i, \varphi_c^d, \varphi_r^d\right)$ generating L such that $L = L(G_T)$ as follows:

- $V = N_H \cup \left(\bigcup_{i=1}^{k} N_i\right) \cup T' \cup \{\#\}$;
- $T = T'$;
- $M = \left\{ \begin{matrix} \# & \# & \# \\ \# & S & \# \end{matrix} \middle| S \in N_H \text{ is the start symbol} \right\}$;
- $C = \left\{ \begin{matrix} \# \\ \alpha \end{matrix} \middle| S \to \alpha \in P_H, \alpha \in (N_H \cup I)^+ \right\} \cup \left\{ \begin{matrix} \# \\ S \end{matrix} \middle| S \to \alpha \in P_H \right\} \cup$

$\left\{ \begin{matrix} \# \\ \gamma \end{matrix} \middle| \beta \to \gamma \in P_H, |\beta| \leq |\gamma| \right\} \cup \left\{ \begin{matrix} \# \\ \beta \end{matrix} \middle| \beta \to \gamma \in P_H, |\beta| \leq |\gamma| \right\}$;

- $R = \left\{ \begin{array}{c} \# \ b \\ \# \ B' \end{array} \middle| B \to bB' \in P_{N_i}, B' \in N_i \right\} \bigcup \left\{ \begin{array}{c} \# \ b \\ \# \ \# \end{array} \middle| B \to b \in P_{T_i} \right\} \bigcup$

$\left\{ \begin{array}{c} b \ \ c \\ B' \ C' \end{array} \middle| B \to bB', C \to cC' \in t_j \subset \bigcup_{i=1}^{k} P_{N_i} \right\} \bigcup$

$\left\{ \begin{array}{c} b \ \ c \\ \# \ \# \end{array} \middle| B \to b, C \to c \in t_j \subset \bigcup_{i=1}^{k} P_{T_i} \right\} \bigcup \left\{ \begin{array}{c} b \ \ \# \\ B' \ \# \end{array} \middle| B \to bB' \in P_{N_i} \right\} \bigcup$

$\left\{ \begin{array}{c} b \ \ \# \\ \# \ \# \end{array} \middle| B \to b \in P_{T_i} \right\} \bigcup \left\{ \# \ B \middle| B \to bB' \in P_{N_i} \right\} \bigcup$

$\left\{ \# \ B \middle| B \to b \in P_{T_i} \right\} \bigcup \left\{ B \ C \middle| B \to bB', C \to cC' \in t_j \subset \bigcup_{i=1}^{k} P_{N_i} \right\} \bigcup$

$\left\{ B \ \# \middle| B \to bB' \in P_{N_i} \right\} \bigcup \left\{ B \ \# \middle| B \to b \in P_{T_i} \right\} (1 \le j \le n).$

We define sets of column and row insertion rules of PCAIDG G corresponding to the rules of TCSMG G_T.

$$\text{IR1} = \left\{ \varphi_c^i \begin{bmatrix} \# & \# \\ \# & S \end{bmatrix} = \left\{ \begin{array}{c} \# \\ \alpha \end{array} \middle| S \to \alpha \in P_H, \alpha \in (N_H \cup I)^+ \right\} \right\}$$

$$\text{DR1} = \left\{ \varphi_c^d \begin{bmatrix} \# & \# \\ \alpha & \# \end{bmatrix} = \left\{ \begin{array}{c} \# \\ S \end{array} \middle| S \to \alpha \in P_H \right\} \right\}$$

$$\text{IR2} = \left\{ \varphi_c^i \begin{bmatrix} \# & \# \\ \# & \beta \end{bmatrix} = \left\{ \begin{array}{c} \# \\ \gamma \end{array} \middle| \beta \to \gamma \in P_H, |\beta| \le |\gamma| \right\} \right\}.$$

Recall if $\alpha = \alpha_1 \alpha_2 \ldots \alpha_l$, then $\begin{array}{c} \# \\ \alpha \end{array} = \begin{array}{cccc} \# & \# & \ldots & \# \\ \alpha_1 & \alpha_2 & \ldots & \alpha_l \end{array}$ where $|\alpha_i| = 1$

$$\text{DR2} = \varphi_c^d \begin{bmatrix} \# & \# \\ \gamma & \# \end{bmatrix} = \left\{ \begin{array}{c} \# \\ \beta \end{array} \middle| \beta \to \gamma \in P_H, |\beta| \le |\gamma| \right\} \Bigg\}.$$

For all $A \in N_H \cup I$, we have

$$\text{IR3} = \left\{ \varphi_c^i \begin{bmatrix} \# & \# \\ A & \beta \end{bmatrix} = \left\{ \begin{array}{c} \# \\ \gamma \end{array} \middle| \beta \to \gamma \in P_H, |\beta| \le |\gamma| \right\} \right\}.$$

$$\text{DR3} = \left\{ \varphi_c^d \begin{bmatrix} \# & \# \\ \gamma & A \end{bmatrix} = \left\{ \begin{array}{c} \# \\ \beta \end{array} \middle| \beta \to \gamma \in P_H, |\beta| \le |\gamma| \right\} \right\}.$$

For every rule of the form $B \to bB'$ in an arbitrary non-terminal table $t_j \subset \bigcup_{i=1}^{k} P_{N_i}$ of G_T, we define corresponding sets of insertion and deletion rules of PCAIDG as follows:

$$\text{IR4} = \left\{ \varphi_r^i \left[\# \, \#, \# \, B \right] = \left\{ \begin{array}{c} \# \ \ b \\ \# \ B' \end{array} \middle| B \to bB' \right\}, \right.$$

$$\varphi_r^i \left[\# \, \#, B \, C \right] = \left\{ \begin{array}{c} b \ \ c \\ B' \ C' \end{array} \middle| B \to bB', C \to cC' \in t_j \right\},$$

$$\left. \varphi_r^i \left[\# \, \#, B \, \# \right] = \left\{ \begin{array}{c} b \ \ \# \\ B' \ \# \end{array} \middle| B \to bB' \right\} \right\}.$$

$$\text{DR4} = \left\{ \varphi_r^d \left[\begin{matrix} \# & b \\ \# & B' \end{matrix}, \lambda\,\lambda \right] = \left\{ \# \, B \,\middle|\, B \to bB' \right\}, \right.$$

$$\varphi_r^d \left[\begin{matrix} b & c \\ B' & C' \end{matrix}, \lambda\,\lambda \right] = \left\{ B \, C \,\middle|\, B \to bB', C \to cC' \in t_j \right\},$$

$$\left. \varphi_r^d \left[\begin{matrix} b & \# \\ B' & \# \end{matrix}, \lambda\,\lambda \right] = \left\{ B \, \# \,\middle|\, B \to bB' \right\} \right\}.$$

For $d, e \in T'$ we have

$$\text{IR5} = \left\{ \varphi_r^i \left[\# \, d, \# \, B \right] = \left\{ \begin{matrix} \# & b \\ \# & B' \end{matrix} \,\middle|\, B \to bB' \right\}, \right.$$

$$\varphi_r^i \left[d \, e, B \, C \right] = \left\{ \begin{matrix} b & c \\ B' & C' \end{matrix} \,\middle|\, B \to bB', C \to cC' \in t_j \right\},$$

$$\left. \varphi_r^i \left[d \, \#, B \, \# \right] = \left\{ \begin{matrix} b & \# \\ B' & \# \end{matrix} \,\middle|\, B \to bB' \right\} \right\}.$$

DR5 $=$ DR4.

For each rule of the form $B \to b$ in an arbitrary terminal table $t_j \subset \bigcup_{i=1}^{k} P_{T_i}$ of G_T, we define the corresponding sets insertion and deletion rules of PCAIDG G as follows:

$$\text{IR6} = \left\{ \varphi_r^i \left[\# \, \#, \# \, B \right] = \left\{ \begin{matrix} \# & b \\ \# & \# \end{matrix} \,\middle|\, B \to b \right\}, \right.$$

$$\varphi_r^i \left[\# \, \#, B \, C \right] = \left\{ \begin{matrix} b & c \\ \# & \# \end{matrix} \,\middle|\, B \to b, C \to c \in t_j \right\},$$

$$\left. \varphi_r^i \left[\# \, \#, B \, \# \right] = \left\{ \begin{matrix} b & \# \\ \# & \# \end{matrix} \,\middle|\, B \to b \right\} \right\}.$$

$$\text{DR6} = \left\{ \varphi_r^d \left[\begin{matrix} \# & b \\ \# & \# \end{matrix}, \lambda\,\lambda \right] = \left\{ \# \, B \,\middle|\, B \to b \right\}, \right.$$

$$\varphi_r^d \left[\begin{matrix} b & c \\ \# & \# \end{matrix}, \lambda\,\lambda \right] = \left\{ B \, C \,\middle|\, B \to b, C \to c \in t_j \right\},$$

$$\left. \varphi_r^d \left[\begin{matrix} b & \# \\ \# & \# \end{matrix}, \lambda\,\lambda \right] = \left\{ B \, \# \,\middle|\, B \to b \right\} \right\}.$$

For $d, e \in T'$, we have

$$\text{IR7} = \left\{ \varphi_r^i \left[\# \, d, \# \, B \right] = \left\{ \begin{matrix} \# & b \\ \# & \# \end{matrix} \,\middle|\, B \to b \right\}, \right.$$

$$\varphi_r^i \left[d \, e, B \, C \right] = \left\{ \begin{matrix} b & c \\ \# & \# \end{matrix} \,\middle|\, B \to b, C \to c \in t_j \right\},$$

$$\left. \varphi_r^i \left[d \, \#, B \, \# \right] = \left\{ \begin{matrix} b & \# \\ \# & \# \end{matrix} \,\middle|\, B \to b \right\} \right\}.$$

DR7 $=$ DR6.

For all $a, b \in T'$, we have

$$\text{DR8} = \left\{ \varphi_c^d \left[\begin{matrix} \lambda & \# \\ \lambda & a \end{matrix} \right] = \left[\begin{matrix} \# \\ \# \end{matrix} \right], \varphi_c^d \left[\begin{matrix} \lambda & a \\ \lambda & b \end{matrix} \right] = \left[\begin{matrix} \# \\ \# \end{matrix} \right], \varphi_c^d \left[\begin{matrix} \lambda & a \\ \lambda & \# \end{matrix} \right] = \left[\begin{matrix} \# \\ \# \end{matrix} \right] \right\},$$

$$\text{DR9} = \left\{ \varphi_c^d \begin{bmatrix} \# & \lambda \\ a & , \lambda \end{bmatrix} = \begin{bmatrix} \# \\ \# \end{bmatrix}, \varphi_c^d \begin{bmatrix} a & \lambda \\ b & , \lambda \end{bmatrix} = \begin{bmatrix} \# \\ \# \end{bmatrix}, \varphi_c^d \begin{bmatrix} a & \lambda \\ \# & , \lambda \end{bmatrix} = \begin{bmatrix} \# \\ \# \end{bmatrix} \right\}.$$

$$\text{DR10} = \left\{ \varphi_r^d \left[\lambda\, \lambda,\, a\, b \right] = \left[\#\, \# \right] \right\}.$$

$$\text{DR11} = \left\{ \varphi_r^d \left[a\, b,\, \lambda\, \lambda \right] = \left[\#\, \# \right] \right\}.$$

□

It is to be noted that sometimes during vertical derivations, rules involving different terminals and non-terminals may end up with same effect and this is attributed to the following reasons:

(i) To simulate the effect of any vertical derivation rule of TCSMG G_T, corresponding contextual rules of PCAIDG G are to be applied more than once depending upon the contexts and
(ii) There are some vertical derivation rules of TCSMG G_T which may get repeated in different tables.

Theorem 2. $\mathscr{L}(TCSMG) \subsetneq \mathscr{L}(PCAIDPS_2)$.

The idea of the proof is as follows:
We show that for every TCSMG generating a language L we can construct a PCAIDPS with 2 membranes generating the same language L. In the PCAIDPS$_2$ that we construct, the skin membrane is the output membrane. We start with the axiom $\begin{matrix} \# & \# & \# \\ \# & S & \# \end{matrix}$ in membrane 2 which corresponds to the start symbol S in the TCSMG. For each horizontal rule in the TCSMG we define one insertion rule and one deletion rule in membrane 2 of PCAIDPS which have the same effect of the rule considered. Similarly, for each vertical rule in an arbitrary table of TCSMG we accordingly define sets of insertion and deletion rules in membrane 2 of PCAIDPS having the same effect as the vertical rule considered. Finally we define deletion rules in PCAIDPS in both membrane 1 and membrane 2 to delete all the remaining #'s in the pictures generated.

Proof. For every TCSMG $G_T = (G_H, G_V)$ with $G_H = (N_H, I, P_H, S)$ and $G_V = \left(\bigcup_{i=1}^{k} G_i, \mathscr{P}\right)$ where $G_i = (N_i, T', P_i, S_i)$, $i = 1, \ldots, k$ are the right-linear grammars and $N_i \cap N_j = \emptyset$ if $i \neq j$, we construct a PCAIDPS$_2$ $\prod = (V, T, [_1[_2]_2]_1, C, R, (M_1, I_1, D_1), (M_2, I_2, D_2), \varphi_c^i, \varphi_r^i, \varphi_c^d, \varphi_r^d, i_0)$ such that $L(\prod) = L(G)$ where

- $V = N_H \bigcup_{i=1}^{k} N_i \bigcup T' \bigcup \{\#\}$;
- $T = T'$;
- $M_1 = \emptyset$;
- $M_2 = \left\{ \begin{matrix} \# & \# & \# \\ \# & S & \# \end{matrix} \right\}$;

- $C = \left\{ \begin{array}{c|c} \# & S \to \alpha \in P_H, \alpha \in (N_H \cup I)^+ \\ \alpha & \end{array} \right\} \bigcup \left\{ \begin{array}{c|c} \# & S \to \alpha \in P_H \\ S & \end{array} \right\} \bigcup$
$\left\{ \begin{array}{c|c} \# & \beta \to \gamma \in P_H, |\beta| \leq |\gamma| \\ \gamma & \end{array} \right\} \bigcup \left\{ \begin{array}{c|c} \# & \beta \to \gamma \in P_H, |\beta| \leq |\gamma| \\ \beta & \end{array} \right\}$;
- $R = \left\{ \begin{array}{c|c} \#\ b & B \to bB' \in P_{N_i}, B' \in N_i \\ \#\ B' & \end{array} \right\} \bigcup \left\{ \begin{array}{c|c} \#\ b & B \to b \in P_{T_i} \\ \#\ \# & \end{array} \right\} \bigcup$
$\left\{ \begin{array}{c|c} b\ \ c & B \to bB', C \to cC' \in t_j \subset \bigcup_{i=1}^k P_{N_i} \\ B'\ C' & \end{array} \right\} \bigcup$
$\left\{ \begin{array}{c|c} b\ \ c & B \to b, C \to c \in t_j \subset \bigcup_{i=1}^k P_{T_i} \\ \#\ \# & \end{array} \right\} \bigcup \left\{ \begin{array}{c|c} b\ \ \# & B \to bB' \in P_{N_i} \\ B'\ \# & \end{array} \right\} \bigcup$
$\left\{ \begin{array}{c|c} b\ \ \# & B \to b \in P_{T_i} \\ \#\ \# & \end{array} \right\} \bigcup \left\{ \#\ B \,\middle|\, B \to bB' \in P_{N_i} \right\} \bigcup$
$\left\{ \#\ B \,\middle|\, B \to b \in P_{T_i} \right\} \bigcup \left\{ B\ C \,\middle|\, B \to bB', C \to cC' \in t_j \subset \bigcup_{i=1}^k P_{N_i} \right\} \bigcup$
$\left\{ B\ \# \,\middle|\, B \to bB' \in P_{N_i} \right\} \bigcup \left\{ B\ \# \,\middle|\, B \to b \in P_{T_i} \right\}$;
- $I_1 = \emptyset$;

Set of deletion rules of D_1 is as follows:

For all $a, b \in T$, we have

$$\text{DR9} = \left\{ \left\{ \varphi_r^d \left[a\, b,\, \lambda\, \lambda \right] = \left[\#\, \# \right] \right\}, here \right\}.$$

Sets of insertion rules of I_2 are defined as follows:

$$\text{IR1} = \left\{ \left\{ \varphi_c^i \begin{bmatrix} \# & \# \\ \# & S \end{bmatrix} = \left\{ \begin{array}{c|c} \# & S \to \alpha \in P_H, \alpha \in (N_H \cup I)^+ \\ \alpha & \end{array} \right\} \right\}, here \right\}$$

$$\text{IR2} = \left\{ \left\{ \varphi_c^i \begin{bmatrix} \# & \# \\ \# & \beta \end{bmatrix} = \left\{ \begin{array}{c|c} \# & \beta \to \gamma \in P_H, |\beta| \leq |\gamma| \\ \gamma & \end{array} \right\} \right\}, here \right\}.$$

For all $A \in (N_H \cup I)$, we have

$$\text{IR3} = \left\{ \left\{ \varphi_c^i \begin{bmatrix} \# & \# \\ A & \beta \end{bmatrix} = \left\{ \begin{array}{c|c} \# & \beta \to \gamma \in P_H, |\beta| \leq |\gamma| \\ \gamma & \end{array} \right\} \right\}, here \right\}.$$

For each non-terminal rule of the form $B \to bB'$ in an arbitrary table $t_j \subset \bigcup_{i=1}^k P_{N_i}$ of G_T, we define corresponding sets of insertion rules of $\prod$ as follows:

$$\text{IR4} = \Bigg\{\Bigg\{\varphi_r^i\left[\#\,\#,\#\,B\right] = \left\{\begin{matrix}\#\ b\\ \#\ B'\end{matrix}\middle| B \to bB\right\},$$
$$\varphi_r^i\left[\#\,\#,B\,C\right] = \left\{\begin{matrix}b\ \ c\\ B'\ C'\end{matrix}\middle| B \to bB', C \to cC' \in t_j\right\},$$
$$\varphi_r^i\left[\#\,\#,B\,\#\right] = \left\{\begin{matrix}b\ \ \#\\ B'\ \#\end{matrix}\middle| B \to bB'\right\}\Bigg\}, here\Bigg\}.$$

For $d, e \in T$, we have

$$\text{IR5} = \Bigg\{\Bigg\{\varphi_r^i\left[\#\,d,\#\,B\right] = \left\{\begin{matrix}\#\ b\\ \#\ B'\end{matrix}\middle| B \to bB'\right\},$$
$$\varphi_r^i\left[d\,e,B\,C\right] = \left\{\begin{matrix}b\ \ c\\ B'\ C'\end{matrix}\middle| B \to bB', C \to cC' \in t_j\right\},$$
$$\varphi_r^i\left[d\,\#,B\,\#\right] = \left\{\begin{matrix}b\ \ \#\\ B'\ \#\end{matrix}\middle| B \to bB'\right\}\Bigg\}, here\Bigg\}.$$

For each rule of the form $B \to b$ in an arbitrary terminal table $t_j \subset \bigcup_{i=1}^{k} P_{T_i}$ of G_T, we define the corresponding sets of insertion rules of $\prod$ as follows:

$$\text{IR6} = \Bigg\{\Bigg\{\varphi_r^i\left[\#\,\#,\#\,B\right] = \left\{\begin{matrix}\#\ b\\ \#\ \#\end{matrix}\middle| B \to b\right\},$$
$$\varphi_r^i\left[\#\,\#,B\,C\right] = \left\{\begin{matrix}b\ \ c\\ \#\ \#\end{matrix}\middle| B \to b, C \to c \in t_j\right\},$$
$$\varphi_r^i\left[\#\,\#,B\,\#\right] = \left\{\begin{matrix}b\ \ \#\\ \#\ \#\end{matrix}\middle| B \to b\right\}\Bigg\}, here\Bigg\}.$$

For $d, e \in T$, we have

$$\text{IR7} = \Bigg\{\Bigg\{\varphi_r^i\left[\#\,d,\#\,B\right] = \left\{\begin{matrix}\#\ b\\ \#\ \#\end{matrix}\middle| B \to b\right\},$$
$$\varphi_r^i\left[d\,e,B\,C\right] = \left\{\begin{matrix}b\ \ c\\ \#\ \#\end{matrix}\middle| B \to b, C \to c \in t_j\right\},$$
$$\varphi_r^i\left[d\,\#,B\,\#\right] = \left\{\begin{matrix}b\ \ \#\\ \#\ \#\end{matrix}\middle| B \to b\right\}\Bigg\}, here\Bigg\}.$$

Sets of deletion rules of the membrane M_2 are defined in D_2 as follows:

$$\text{DR1} = \Bigg\{\Bigg\{\varphi_c^d\begin{bmatrix}\#\\ \alpha\end{bmatrix}, \begin{bmatrix}\#\\ \#\end{bmatrix} = \left\{\begin{matrix}\#\\ S\end{matrix}\middle| S \to \alpha \in P_H\right\}\Bigg\}, here\Bigg\}$$
$$\text{DR2} = \Bigg\{\Bigg\{\varphi_c^d\begin{bmatrix}\#\\ \gamma\end{bmatrix}, \begin{bmatrix}\#\\ \#\end{bmatrix} = \left\{\begin{matrix}\#\\ \beta\end{matrix}\middle| \beta \to \gamma \in P_H, |\beta| \leq |\gamma|\right\}\Bigg\}, here\Bigg\}.$$

For all $A \in (N_H \cup I)$, we have

$$\text{DR3} = \left\{\left\{\varphi_c^d\left[\begin{matrix}\# \\ \gamma\end{matrix}, \begin{matrix}\# \\ A\end{matrix}\right] = \left\{\begin{matrix}\# \\ \beta\end{matrix}\middle| \beta \to \gamma \in P_H, |\beta| \le |\gamma|\right\}\right\}, here\right\}.$$

For each rule of the form $B \to bB'$ in an arbitrary non-terminal table $t_j \subset \bigcup_{i=1}^{k} P_{N_i}$ of G_T, we define the corresponding sets of deletion rules of $\prod$ as follows:

$$\text{DR4} = \left\{\left\{\varphi_r^d\left[\begin{matrix}\# & b \\ \# & B'\end{matrix}, \lambda\,\lambda\right] = \left\{\#\, B \middle| B \to bB'\right\},\right.\right.$$

$$\varphi_r^d\left[\begin{matrix}b & c \\ B' & C'\end{matrix}, \lambda\,\lambda\right] = \left\{B\, C \middle| B \to bB', C \to cC' \in t_j\right\},$$

$$\left.\left.\varphi_r^d\left[\begin{matrix}b & \# \\ B' & \#\end{matrix}, \lambda\,\lambda\right] = \left\{B\, \# \middle| B \to bB'\right\}\right\}, here\right\}.$$

For each rule of the form $B \to b$ in an arbitrary terminal table $t_j \subset \bigcup_{i=1}^{k} P_{T_i}$ of G_T, we define the corresponding sets of deletion rules in $\prod$ as follows:

$$\text{DR5} = \left\{\left\{\varphi_r^d\left[\begin{matrix}\# & b \\ \# & \#\end{matrix}, \lambda\,\lambda\right] = \left\{\#\, B \middle| B \to b\right\},\right.\right.$$

$$\varphi_r^d\left[\begin{matrix}b & c \\ \# & \#\end{matrix}, \lambda\,\lambda\right] = \left\{B\, C \middle| B \to b, C \to c \in t_j\right\},$$

$$\left.\left.\varphi_r^d\left[\begin{matrix}b & \# \\ \# & \#\end{matrix}, \lambda\,\lambda\right] = \left\{B\, \# \middle| B \to b\right\}\right\}, here\right\}.$$

For all $a, b \in T$, we have

$$\text{DR6} = \left\{\left\{\varphi_c^d\left[\begin{matrix}\lambda \\ \lambda\end{matrix}, \begin{matrix}\# \\ a\end{matrix}\right] = \left[\begin{matrix}\# \\ \#\end{matrix}\right], \varphi_c^d\left[\begin{matrix}\lambda \\ \lambda\end{matrix}, \begin{matrix}a \\ b\end{matrix}\right] = \left[\begin{matrix}\# \\ \#\end{matrix}\right],\right.\right.$$

$$\left.\left.\varphi_c^d\left[\begin{matrix}\lambda \\ \lambda\end{matrix}, \begin{matrix}a \\ \#\end{matrix}\right] = \left[\begin{matrix}\# \\ \#\end{matrix}\right]\right\}, here\right\}.$$

For all $a, b \in T$, we have

$$\text{DR7} = \left\{\left\{\varphi_c^d\left[\begin{matrix}\# \\ a\end{matrix}, \begin{matrix}\lambda \\ \lambda\end{matrix}\right] = \left[\begin{matrix}\# \\ \#\end{matrix}\right], \varphi_c^d\left[\begin{matrix}a \\ b\end{matrix}, \begin{matrix}\lambda \\ \lambda\end{matrix}\right] = \left[\begin{matrix}\# \\ \#\end{matrix}\right],\right.\right.$$

$$\left.\left.\varphi_c^d\left[\begin{matrix}a \\ \#\end{matrix}, \begin{matrix}\lambda \\ \lambda\end{matrix}\right] = \left[\begin{matrix}\# \\ \#\end{matrix}\right]\right\}, here\right\}.$$

$$\text{DR8} = \left\{\left\{\varphi_r^d\left[\lambda\,\lambda, a\,b\right] = \left[\#\,\#\right]\right\}, out\right\}$$

Working Procedure of P System: In membrane 2, there are seven sets of column and row insertion rules of I_2 grouped according to the rules of G_H and G_V respectively of the grammar G_T.

For every rule $S \to \alpha \in P$ in G_H, column insertion rules to insert the context $\begin{smallmatrix}\#\\ \alpha\end{smallmatrix}$ between $\begin{smallmatrix}\#\\ \#\end{smallmatrix}$ and $\begin{smallmatrix}\#\\ S\end{smallmatrix}$ are defined.

For every rule of the form $\beta \to \gamma \in P$ in G_H, the corresponding column insertion rules are defined to insert the context $\begin{smallmatrix}\#\\ \gamma\end{smallmatrix}$ between $\begin{smallmatrix}\#\\ \#\end{smallmatrix}$ and $\begin{smallmatrix}\#\\ \beta\end{smallmatrix}$ (or) $\begin{smallmatrix}\#\\ A\end{smallmatrix}$ and $\begin{smallmatrix}\#\\ \beta\end{smallmatrix}$.

In membrane 2, there are eight sets of column and row deletion rules defined in D_2 based on the rules of G_H and G_V respectively of the grammar G_T.

For every rule $S \to \alpha \in P$ in G_H, deletion rules to delete the contexts $\begin{smallmatrix}\#\\ S\end{smallmatrix}$ between $\begin{smallmatrix}\#\\ \alpha\end{smallmatrix}$ and $\begin{smallmatrix}\#\\ \#\end{smallmatrix}$ are defined.

For every rule of the form $\beta \to \gamma \in P$ in G_H, the corresponding column deletion rules are defined to delete the context $\begin{smallmatrix}\#\\ \beta\end{smallmatrix}$ between $\begin{smallmatrix}\#\\ \alpha\end{smallmatrix}$ and $\begin{smallmatrix}\#\\ \#\end{smallmatrix}$ (or) $\begin{smallmatrix}\#\\ \gamma\end{smallmatrix}$ and $\begin{smallmatrix}\#\\ \#\end{smallmatrix}$.

Using these column insertion and deletion rules of the P system, the results effected by horizontal derivation of the horizontal grammar G_H of G_T can be achieved. It is to be noted that the rules for the horizontal growth in I_2 and D_2 of the P system are defined in such a manner that the column insertion and deletion rules are applied alternatively.

To simulate the vertical derivation of G_T, row insertion and deletion rules are defined based on the rules of the vertical grammars of G_V.

For each non-terminal rule of the form $B \to bB'$ in an arbitrary table $t_j \subset \bigcup_{i=1}^{k} P_{N_i}$ of G_T, the corresponding row insertion and row deletion rules are defined in I_2 and D_2 respectively to replicate the vertical derivation of the vertical grammars G_V.

It is again to be noted that the rules for vertical growth in I_2 and D_2 of the P system are defined in such a manner that the row insertion and row deletion rules are applied alternatively. A row insertion rule in I_2 is defined to send the generated picture to the skin membrane which is the output membrane.

The working of the P system in membrane 2 is as follows: The axiom set consists of the array $\begin{matrix}\# & \# & \#\\ \# & S & \#\end{matrix}$ based on the starting symbol S of any TCSMG G. We consider the rules in I_2 and D_2 to perform the parallel contextual column insertion and column deletion operations and these operations are performed alternatively to simulate the generation of horizontal strings of intermediates based on G_H. Now we consider the rules in I_2 and D_2 to perform the parallel contextual row insertion operation and row deletion operation. Parallel contextual row insertion and deletion operations are performed alternatively to simulate the vertical generation of the picture based on G_V. Then using the parallel contextual column deletion rules in D_2, the #'s along the borders of the columns are deleted. By using the parallel contextual row deletion rule in D_2, the #'s along the top border of the arrays are deleted and the resulting arrays are sent out of the membrane 2.

Finally, using the parallel contextual row deletion rules in D_1, the #'s along the bottom border of the arrays are deleted and the resulting pictures belonging to $L(G_T)$ are collected in membrane 1, the output membrane. □

A Sample Computation:

We now exhibit a sample computation of a picture of size 3×8 of $L(G_2)$ of Example 2 using the PCAIDPS_2 $\prod_3$ that can be constructed as per Theorem 2 from G_2 as follows:

$$\begin{array}{ccc} \# & \# & \# \\ \# & S & \# \end{array} \Rightarrow^{col_i}_{(2,IR1)} \begin{array}{ccccccc} \# & \# & \# & \# & \# & \# & \# \\ \# & S_1 & S & C & S_1 & S & \# \end{array} = \begin{array}{ccccccc} \# & \# & \# & \# & \# & \# & \# \\ \# & S_1 & S & C & S_1 & S & \# \end{array} \Rightarrow^{col_d}_{(2,DR1)}$$

$$\begin{array}{cccccc} \# & \# & \# & \# & \# & \# \\ \# & S_1 & S & C & S_1 & \# \end{array} \Rightarrow^{col_i}_{(2,IR2)} \begin{array}{ccccccccccc} \# & \# & \# & \# & \# & \# & \# & \# & \# & \# & \# \\ \# & S_1 & S_1 & S_2 & S_1 & S_2 & S_1 & S & C & S_1 & \# \end{array} =$$

$$\begin{array}{ccccccccccc} \# & \# & \# & \# & \# & \# & \# & \# & \# & \# & \# \\ \# & S_1 & S_1 & S_2 & S_1 & S_2 & S_1 & S & C & S_1 & \# \end{array} \Rightarrow^{col_d}_{(2,DR2)}$$

$$\begin{array}{cccccccccc} \# & \# & \# & \# & \# & \# & \# & \# & \# & \# \\ \# & S_1 & S_1 & S_2 & S_1 & S_2 & S_1 & C & S_1 & \# \end{array} \Rightarrow^{col_i}_{(2,IR2)}$$

$$\begin{array}{ccccccccccc} \# & \# & \# & \# & \# & \# & \# & \# & \# & \# & \# \\ \# & S_1 & S_1 & S_2 & S_1 & S_2 & C & S_1 & S_1 & C & S_1 & \# \end{array} =$$

$$\begin{array}{ccccccccccc} \# & \# & \# & \# & \# & \# & \# & \# & \# & \# & \# \\ \# & S_1 & S_1 & S_2 & S_1 & S_2 & C & S_1 & S_1 & C & S_1 & \# \end{array} \Rightarrow^{col_d}_{(2,DR2)}$$

$$\begin{array}{cccccccccc} \# & \# & \# & \# & \# & \# & \# & \# & \# & \# \\ \# & S_1 & S_1 & S_2 & S_1 & S_2 & C & S_1 & S_1 & \# \end{array} \Rightarrow^{col_i}_{(2,IR2)}$$

$$\begin{array}{ccccccccccc} \# & \# & \# & \# & \# & \# & \# & \# & \# & \# & \# \\ \# & S_1 & S_1 & S_2 & S_1 & S_1 & S_2 & S_2 & C & S_1 & S_1 & \# \end{array} =$$

$$\begin{array}{ccccccccccc} \# & \# & \# & \# & \# & \# & \# & \# & \# & \# & \# \\ \# & S_1 & S_1 & S_2 & S_1 & S_1 & S_2 & S_2 & C & S_1 & S_1 & \# \end{array} \Rightarrow^{col_d}_{(2,DR2)}$$

$$\begin{array}{cccccccccc} \# & \# & \# & \# & \# & \# & \# & \# & \# & \# \\ \# & S_1 & S_1 & S_2 & S_1 & S_1 & S_2 & S_1 & S_1 & \# \end{array} \Rightarrow^{row_i}_{(2,IR3)} \begin{array}{cccccccccc} \# & \# & \# & \# & \# & \# & \# & \# & \# & \# \\ \# & x & x & \bullet & x & x & \bullet & x & x & \# \\ \# & S_1 & S_1 & A & S_1 & S_1 & A & S_1 & S_1 & \# \\ \# & S_1 & S_1 & S_2 & S_1 & S_1 & S_2 & S_1 & S_1 & \# \end{array} =$$

$$\begin{array}{cccccccccc} \# & \# & \# & \# & \# & \# & \# & \# & \# & \# \\ \# & x & x & \bullet & x & x & \bullet & x & x & \# \\ \# & S_1 & S_1 & A & S_1 & S_1 & A & S_1 & S_1 & \# \\ \# & S_1 & S_1 & S_2 & S_1 & S_1 & S_2 & S_1 & S_1 & \# \end{array} \Rightarrow^{row_d}_{(2,DR4)}$$

$$\begin{matrix} \# & \# & \# & \# & \# & \# & \# & \# & \# & \# \\ \# & x & x & \bullet & x & x & \bullet & x & x & \# \\ \# & S_1 & S_1 & A & S_1 & S_1 & A & S_1 & S_1 & \# \end{matrix} \Rightarrow^{row_i}_{(2,IR4)} \begin{matrix} \# & \# & \# & \# & \# & \# & \# & \# & \# & \# \\ \# & x & x & \bullet & x & x & \bullet & x & x & \# \\ \# & x & x & x & x & x & x & x & x & \# \\ \# & S_1 & S_1 & B & S_1 & S_1 & B & S_1 & S_1 & \# \\ \# & S_1 & S_1 & A & S_1 & S_1 & A & S_1 & S_1 & \# \end{matrix} =$$

$$\begin{matrix} \# & \# & \# & \# & \# & \# & \# & \# & \# & \# \\ \# & x & x & \bullet & x & x & \bullet & x & x & \# \\ \# & x & x & x & x & x & x & x & x & \# \\ \# & S_1 & S_1 & B & S_1 & S_1 & B & S_1 & S_1 & \# \\ \# & S_1 & S_1 & A & S_1 & S_1 & A & S_1 & S_1 & \# \end{matrix} \Rightarrow^{row_d}_{(2,DR4)} \begin{matrix} \# & \# & \# & \# & \# & \# & \# & \# & \# & \# \\ \# & x & x & \bullet & x & x & \bullet & x & x & \# \\ \# & x & x & x & x & x & x & x & x & \# \\ \# & S_1 & S_1 & B & S_1 & S_1 & B & S_1 & S_1 & \# \end{matrix} \Rightarrow^{row_i}_{(2,IR5)}$$

$$\begin{matrix} \# & \# & \# & \# & \# & \# & \# & \# & \# & \# \\ \# & x & x & \bullet & x & x & \bullet & x & x & \# \\ \# & x & x & x & x & x & x & x & x & \# \\ \# & x & x & \bullet & x & x & \bullet & x & x & \# \\ \# & \# & \# & \# & \# & \# & \# & \# & \# & \# \\ \# & S_1 & S_1 & B & S_1 & S_1 & B & S_1 & S_1 & \# \end{matrix} = \begin{matrix} \# & \# & \# & \# & \# & \# & \# & \# & \# & \# \\ \# & x & x & \bullet & x & x & \bullet & x & x & \# \\ \# & x & x & x & x & x & x & x & x & \# \\ \# & x & x & \bullet & x & x & \bullet & x & x & \# \\ \# & \# & \# & \# & \# & \# & \# & \# & \# & \# \\ \# & S_1 & S_1 & B & S_1 & S_1 & B & S_1 & S_1 & \# \end{matrix} \Rightarrow^{row_d}_{(2,DR4)}$$

$$\begin{matrix} \# & \# & \# & \# & \# & \# & \# & \# & \# & \# \\ \# & x & x & \bullet & x & x & \bullet & x & x & \# \\ \# & x & x & x & x & x & x & x & x & \# \\ \# & x & x & \bullet & x & x & \bullet & x & x & \# \\ \# & \# & \# & \# & \# & \# & \# & \# & \# & \# \end{matrix} \Rightarrow^{col_d}_{(2,DR5)} \begin{matrix} \# & \# & \# & \# & \# & \# & \# & \# & \# \\ x & x & \bullet & x & x & \bullet & x & x & \# \\ x & x & x & x & x & x & x & x & \# \\ x & x & \bullet & x & x & \bullet & x & x & \# \\ \# & \# & \# & \# & \# & \# & \# & \# & \# \end{matrix} \Rightarrow^{col_d}_{(2,DR6)}$$

$$\begin{matrix} \# & \# & \# & \# & \# & \# & \# & \# \\ x & x & \bullet & x & x & \bullet & x & x \\ x & x & x & x & x & x & x & x \\ x & x & \bullet & x & x & \bullet & x & x \\ \# & \# & \# & \# & \# & \# & \# & \# \end{matrix} \Rightarrow^{row_d}_{(1,DR7)} \begin{matrix} x & x & \bullet & x & x & \bullet & x & x \\ x & x & x & x & x & x & x & x \\ x & x & \bullet & x & x & \bullet & x & x \\ \# & \# & \# & \# & \# & \# & \# & \# \end{matrix} \Rightarrow^{row_d}_{(1,DR8)}$$

$$\begin{matrix} x & x & \bullet & x & x & \bullet & x & x \\ x & x & x & x & x & x & x & x. \\ x & x & \bullet & x & x & \bullet & x & x \end{matrix}$$

In the above derivation, we have not used all the rules of $\prod_3$ but only the rules necessary for the derivation of the picture were used. It is to be noted that we have not listed other rules of the computation owing to space constraint.

5 Conclusion

This paper is a step forward in the attempt/journey to bring in a hierarchy among two-dimensional picture languages. Still there are other well known families of two-dimensional picture languages to be compared with the family of languages generated by PCAIDPS and thereby the generative power of PCAIDPS is revealed further and will be taken up in our future work. Some classes of P systems to be compared imminently with PCAIDPS in terms of their generative powers are:

(i) Parallel Contextual Array P Systems of James et al. [14].
(ii) Contextual Array Grammars and Array P Systems of Henning Fernau et al. [8].
(iii) Array rewriting P systems of Rodica Ceterchi et al. [2].

Daniel Díaz-Pernil et al. [6] surveyed applications of P systems in image processing. Rodica Ceterchi et al. [3] have constructed P systems to generate the approximations of geometric patterns of space filling curves such as Peano's curve, Hilbert's curves and others. In this paper, we deal with applications of P systems in terms of rectangular arrays of letters yielding pictures or images by replacing letters by primitive symbols. The role of PCAIDPS in image analysis can be explored further.

References

1. Alhazov, A., Krassovitskiy, A., Rogozhin, Y., Verlan, S.: P systems with minimal insertion and deletion. Theoret. Comput. Sci. **412**(1), 136–144 (2011)
2. Ceterchi, R., Mutyam, M., Paun, G., Subramanian, K.G.: Array-rewriting P systems. Nat. Comput. **2**(3), 229–249 (2003)
3. Ceterchi, R., Subramanian, K.G.: Generating pictures in string representation with P systems: the case of space-filling curves. J. Membrane Comput. **2**(4), 369–379 (2020). https://doi.org/10.1007/s41965-020-00061-z
4. Helen Chandra, P., Subramanian, K.G., Thomas, D.G.: Parallel contextual array grammars and languages. Electron. Notes Discret. Math. **12**, 106–117 (2003)
5. Dersanambika, K.S., Krithivasan, K.: Contextual array P systems. Int. J. Comput. Math. **81**(8), 955–969 (2004)
6. Díaz-Pernil, D., Gutiérrez-Naranjo, M.A., Peng, H.: Membrane computing and image processing: a short survey. J. Membrane Comput. **1**(1), 58–73 (2019). https://doi.org/10.1007/s41965-018-00002-x
7. Domaratzki, M., Okhotin, A.: Representing recursively enumerable languages by iterated deletion. Theor. Comput. Sci. **314**(3), 451–457 (2004)
8. Fernau, H., Freund, R., Schmid, M.L., Subramanian, K.G., Wiederhold, P.: Contextual array grammars and array P systems. Ann. Math. Artif. Intell. (1), 5–26 (2013). https://doi.org/10.1007/s10472-013-9388-0
9. Rudolf, F., Gheorghe, P., Rozenberg, G.: Contextual array grammars. In: Formal Models, Languages and Applications, volume 66 of Series in Machine Perception and Artificial Intelligence, pp. 112–136. World Scientific (2007)
10. Giammarresi, D., Restivo, A.: Recognizable picture languages. Int. J. Pattern Recognit. Artif. Intell. **6**(2&3), 241–256 (1992)
11. Giammarresi, D., Restivo, A.: Two-dimensional languages. In: Rozenberg, G., Salomaa, A. (eds.) Handbook of Formal Languages, pp. 215–267. Springer, Heidelberg (1997). https://doi.org/10.1007/978-3-642-59126-6_4

12. Haussler, D.H.: Insertion and iterated insertion as operations on formal languages. Ph.D. thesis, University of Colorado (1982)
13. Immanuel, S.J., Jayasankar, S., Gnanaraj Thomas, D., Paramasivan, M., Thamburaj, R., Nagar, A.K.: Parallel contextual array insertion deletion P systems and Siromoney matrix grammars. Int. J. Parallel Emergent Distrib. Syst. **36** (2021)
14. Bera, S., Pan, L., Song, B., Subramanian, K.G., Zhang, G.: Parallel contextual array P systems. Int. J. Adv. Eng. Sci. Appl. Math. **10**(3), 203–212 (2018). https://doi.org/10.1007/s12572-018-0226-9
15. James Immanuel, S., Thomas, D.G., Thamburaj, R., Nagar, A.K.: Parallel contextual array insertion deletion P system. In: Brimkov, V.E., Barneva, R.P. (eds.) IWCIA 2017. LNCS, vol. 10256, pp. 170–183. Springer, Cham (2017). https://doi.org/10.1007/978-3-319-59108-7_14
16. Thomas, D.G., Immanuel, S.J., Nagar, A.K., Thamburaj, R.: Parallel contextual array insertion deletion grammar. In: Barneva, R.P., Brimkov, V.E., Tavares, J.M.R.S. (eds.) IWCIA 2018. LNCS, vol. 11255, pp. 28–42. Springer, Cham (2018). https://doi.org/10.1007/978-3-030-05288-1_3
17. Immanuel, S.J., Thomas, D.G.: Parallel contextual array P system with contexts shuffled on trajectories. In: Proceedings of National Conference on Mathematics and Computer Applications, NCMCA 2015, pp. 214–222 (2015)
18. Ito, M., Kari, L., Thierrin, G.: Insertion and deletion closure of languages. Theor. Comput. Sci. **183**(1), 3–19 (1997)
19. Immanuel, S.J., Jayasankar, S., Gnanaraj Thomas, D., Paramasivan, M., Thamburaj, R., Nagar, A.K.: Parallel contextual array insertion deletion P systems and Siromoney matrix grammars. In: Pre-Proceedings of the 8th Asian Conference on Membrane Computing, pp. 134–151 (2019)
20. Lila, K.: On insertion and deletion in formal languages. Ph.D. thesis, University of Turku (1991)
21. Kari, L., Thierrin, G.: Contextual insertions/deletions and computability. Inf. Comput. **131**(1), 47–61 (1996)
22. Krishna, S.N., Rama, R.: Insertion-deletion P systems. In: Jonoska, N., Seeman, N.C. (eds.) DNA 2001. LNCS, vol. 2340, pp. 360–370. Springer, Heidelberg (2002). https://doi.org/10.1007/3-540-48017-X_34
23. Krithivasan, K., Mutyam, M.: Contextual P systems. Fundam. Informaticae **49**(1–3), 179–189 (2002)
24. Krithivasan, K., Siromoney, R.: Characterizations of regular and context-free matrices. Int. J. Comput. Math. **4**(1–4), 229–245 (1974)
25. Marcus, S.: Contextual grammars. Rev. Roum. Math. Pures Appl. **14**, 1525–1534 (1969)
26. Paun, G.: Computing with membranes. J. Comput. Syst. Sci. **61**(1), 108–143 (2000)
27. Paun, Gheorghe: Membrane Computing: An Introduction. Natural Computing Series. Springer, Heidelberg (2002). https://doi.org/10.1007/978-3-642-56196-2
28. Rosebrugh, R.D., Wood, D.: Restricted parallelism and right linear grammars. Utilitas Mathematica **7**, 151–186 (1975)
29. Siromoney, G., Siromoney, R., Krithivasan, K.: Abstract families of matrices and picture languages. Comput. Graph. Image Process. **1**(3), 284–307 (1972)
30. Siromoney, R., Subramanian, K.G., Rangarajan, K.: Parallel/Sequential rectangular arrays with tables. Int. J. Comput. Math. **6**(2), 143–158 (1977)
31. Subramanian, K.G., Van, D.L., Chandra, P.H., Quyen, N.D.: Array grammars with contextual operations. Fundam. Informaticae **83**(4), 411–428 (2008)

Triangular Array Token Petri Net and P System

T. Kalyani[1(✉)], T. T. Raman[1], D. G. Thomas[2], K. Bhuvaneswari[3], and P. Ravichandran[4]

[1] Department of Mathematics, St. Joseph's Institute of Technology, Chennai 119, Tamilnadu, India
[2] Department of Applied Mathematics, Saveetha School of Engineering, SIMATS, Chennai 602 105, Tamilnadu, India
[3] Department of Mathematics, Sathyabama Institute of Science and Technology, Chennai 119, Tamilnadu, India
[4] Department of Mechanical Engineering, St. Joseph's Institute of Technology, Chennai 119, Tamilnadu, India

Abstract. A Petri Net is a mathematical model used to generate string languages, which is useful in data analysis, pattern matchings, simulations etc. Array Token Petri Nets were introduced to generate two-dimensional and three-dimensional picture languages. In this paper, we introduce Triangular Array Token Petri Net (TATPN) to generate certain interesting patterns of triangular picture languages using Elementary Evolution Rules (EER) and Parallel Evolution Rules (PER). We also introduce Triangular Array Token Petri Net P System and compared it with TATPN and TTPPS for generative power.

Keywords: Petri Net · Array-Token Petri Net · Triangular Picture languages · P system

1 Introduction

Petri Nets are mathematical models introduced to model dynamic systems [4, 13]. It is a bipartite graph which has directed arcs connecting place nodes and transition nodes. Input place is a place from which the directed arc starts and the place to which an arc enters from a transition is called the output place of transition. Places may hold any number of tokens. Distribution of tokens over the places of net is called a marking. A transition triggers whenever there is at least one token in all the input places. When a transition fires, tokens are removed from its input places and added to all output places of the transition.

Array-Token Petri Nets are introduced in [5,8,11,14] to generate two-dimensional picture languages. Array-token petri nets to generate hexagonal, Octagonal and triangular picture languages were introduced in [1,7,9]. 3D-Array Token Petri nets generating Tetrahedral Picture Languages is introduced

R. Freund et al. (Eds.): CMC 2020, LNCS 12687, pp. 78–93, 2021.
https://doi.org/10.1007/978-3-030-77102-7_5

in [6]. Petri Net model to generate tiling and kolam patterns were introduced in [10,15].

On the other hand P system or membrane computing, which was introduced by Gh. Paun [12] inspired by the functioning of living cells has applications in different areas, with the area of picture generation being one among these application areas. P system is a distributed, highly parallel theoretical computing model, based on the membrane structure and the behaviour of the living cells. The basic model processes multi-sets of objects in the region that are defined by a hierarchical arrangement of membranes by evolution rules associated with the regions. Several P system models have been proposed in the literature for picture generation [3,16,17]. Among a variety of applications of the P system model, the problem of handling array languages using P systems was considered by Ceterchi et al. [3] by introducing array-rewriting P systems and thus linking the two areas of membrane computing and picture grammars. Triangular tile pasting *P* system (TTPPS) is introduced in [2].

In this paper, we propose Triangular Array Token Petri Net (TATPN) to generate some interesting triangular picture patterns. Triangular Array Token Petri Net P System (TATPNPS) is also introduced and is compared with TATPN and TTPPS for generative power.

2 Preliminaries

In this section, we recall the definitions of isosceles right angled triangular tiles, catenation rules and Array Token Petri Net Structure.

Definition 1. *[1] Let* Σ = {A, B, C, D} *be a finite set of labeled isosceles right angled triangular tiles of dimensions* $\frac{1}{\sqrt{2}}, \frac{1}{\sqrt{2}}$ *and 1 unit obtained by intersecting a unit square by its diagonals. Catenation rules of tile A are as follows: Tiles which can be catenated with A are B, C and D by the rules* $\{(a_1, b_1), (a_2, b_2), (a_3, b_3)\}, \{(a_3, c_1)\}, \{(a_1, d_3)\}$. *Similar catenation rules can be defined for the remaining tiles.*

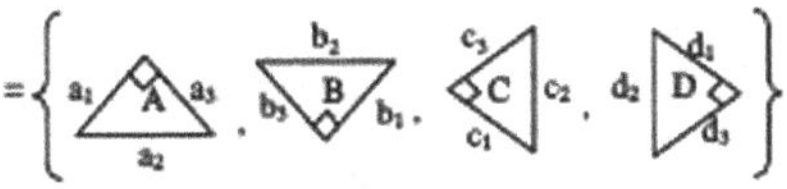

Definition 2. *Let* Γ= {E, F, G, H} *be a finite set of labeled isosceles right angled triangular tiles of dimensions 1, 1 and* $\sqrt{2}$ *unit obtained by intersecting a unit square by its right diagonal to get tiles E and F and by its left diagonal to get tiles G and H. Catenation rules are as follows: Tile E can be catenated with tiles F, G and H by the rules* $\{(e_1, f_1), (e_2, f_2), (e_3, f_3)\}, \{(e_1, g_2)\}, \{(e_2, h_1)\}$. *Similar catenation rules can be defined for the remaining tiles.*

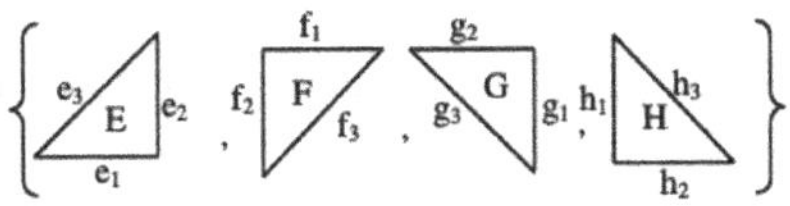

Definition 3. *The tiles of* Σ *can be catenated with the tiles of* Γ *by the following rules:*

(i) Tile A can be catenated with tiles F and G by the rules $\{(a_2, f_1)\}$ *and* $\{(a_2, g_2)\}$.
(ii) Tile B can be catenated with tiles E and H by the rules $\{(b_2, e_1)\}$ *and* $\{(b_2, h_2)\}$.
(iii) Tile C can be catenated with tiles F and H by the rules $\{(c_2, f_2)\}$ *and* $\{(c_2, h_1)\}$.
(iv) Tile D can be catenated with tiles E and G by the rules $\{(d_2, e_2)\}$ *and* $\{(d_2, g_1)\}$.

Definition 4. *Let* $\Sigma^T = \Sigma \cup \Gamma$. *A picture formed by catenating triangular tiles of* Σ^T *along their gluable edges is called a triangular picture or a triangular array.*

Example 1.

Let $\Sigma^T = \{$ A (a_1, a_2, a_3), B (b_1, b_2, b_3), C (c_1, c_2, c_3), D (d_1, d_2, d_3), E (e_1, e_2, e_3), F (f_1, f_2, f_3), G (g_1, g_2, g_3), H (h_1, h_2, h_3) $\}$

A triangular picture formed by catenating these tiles along their gluable edges by the catenation rules $\{(a_3, c_1), (c_3, b_1), (b_1, d_1), (b_2, e_1), (c_2, h_1), (a_2, f_1), (d_2, g_1)\}$ is given in Fig. 1.

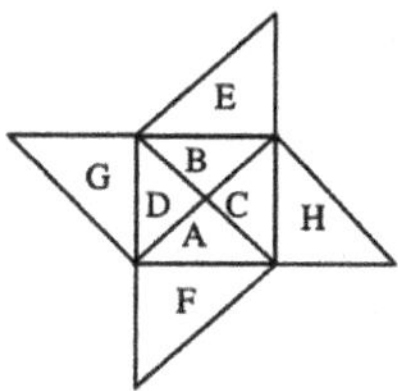

Fig. 1. Triangular picture or triangular array

An Array Token Petri Net (ATPN) generating certain patterns was introduced in [15] which uses Elementary Evolution Rule (EER) and Parallel Evolution Rule (PER). The language, which is generated by ATPN with EER on transition is denoted by AERL.

3 Triangular Array Token Petri Net

In this section, we introduce Triangular Array Token Petri Net (TATPN) with Elementary Evolution Rule (EER) and Parallel Evolution Rule (PER) to generate some interesting triangular picture patterns.

Definition 5. *A Triangular Array Token Petri Net (TATPN) is a 6-tuple* $TN = (P, T, C, A, R, M_0)$ *where* P *is a set of places,* T *is a set of transitions,* C *is a set of symbols (colours) and* C_{AY} *is the set of all triangular arrays over* C*, that are associated with the tokens,* $A \subseteq (P \times T) \cup (T \times P)$ *is a set of arcs,* $R(t)$ *is the set of evolution rules associated with a transition* t*,* M_0 *the initial marking is a function defined on* P *such that for* $p \in P, M_0(p) \in [C_{AY}]_{MS}$.

Definition 6. *An Elementary Evolution Rule over* Σ^T_{AY} *where* Σ^T *is a finite alphabet of triangular tiles*

is one of the following:

(i) *Identity, which keeps the triangular array unaltered.*
(ii) *Horizontal insertion*
  (i) $\lambda \to A$ *or* $B(u/d$*, according as it is up or down)*
  (ii) $\lambda \to E$ *or* $F(u/d$*, according as it is up or down)*
  (iii) $\lambda \to H$ *or* $G(u/d$*, according as it is up or down)*
(iii) *Vertical insertion*
  (i) $\lambda \to C$ *or* $D(l/r$*, according as it is left or right)*
  (ii) $\lambda \to E$ *or* $H(l/r$*, according as it is left or right)*
  (iii) $\lambda \to G$ *or* $F(l/r$*, according as it is left or right)*
(iv) *Right up insertion* $\lambda \to B$ *or* $C(ru)$
(v) *Right down insertion* $\lambda \to A$ *or* $C(rd)$
(vi) *Left up insertion* $\lambda \to B$ *or* $D(lu)$
(vii) *Left down insertion* $\lambda \to A$ *or* $D(ld)$
(viii) *Right down diagonal insertion* $\lambda \to E(rdd)$
(ix) *Right up diagonal insertion* $\lambda \to G(rud)$
(x) *Left down diagonal insertion* $\lambda \to H(ldd)$
(xi) *Left up diagonal insertion* $\lambda \to F(lud)$.

The subnet in Fig. 2 illustrates how horizontal insertion rule is applied to up or down of triangular tile A *or* B *respectively. The rule* $\lambda \to A(u)$*, inserts the triangular tile* A *above tile* B.

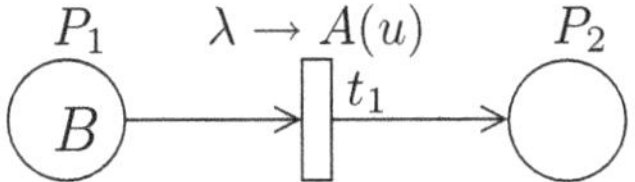

Fig. 2. Subnet used for horizontal insertion

The resultant triangular picture is of the form A B*. In a similar way, it can be done for the other insertion rules.*

The triangular picture language generated by TATPN with EER on transition is denoted by TAERL.

Definition 7. *A Parallel Evolution Rule (PER) over* Σ_{AY}^{T} *where* Σ^{T} *is a finite alphabet of triangular tiles is one of the following:*

(i) *The rule* $\lambda \rightarrow q(u, d, r, l, lu, ld, ru, rd, rdd, rud, lud, ldd)$ *inserts the triangular picture* q *simultaneously on up, down, right, left, left up, left down, right up, right down, right down diagonal, right up diagonal, left up diagonal and left down diagonal provided the triangular picture* q *is gluable with those edges on the given directions.*

(ii) *The rule* $\lambda \rightarrow q_1(u, d); q_2(l, r); q_3(ru, rd); q_4(lu, ld); q_5(ur, r); q_6(rd, d); q_7(lu, l); q_8(ld, l)$; *inserts* q_1 *on up and down provided the edges are gluable. In a similar way, explanations can be given for the other rules.*

(iii) *Let* $c_1, c_2, c_3, c_4, c_5, c_6, c_7, c_8, c_9, c_{10}, c_{11}$ *and* c_{12} *denote up, right up, right, right down, down, left down, left, left up, right up diagonal, right down diagonal, left up diagonal and left down diagonal corners of a triangular picture respectively.*

The rule $\lambda \rightarrow q(c_1, c_2, c_3, c_4, c_5, c_6, c_7, c_8, c_9, c_{10}, c_{11}, c_{12})$ *inserts* q *up, right up, right, right down, down, left down, left, left up, right up diagonal, right down diagonal, left up diagonal and left down diagonal corners of a triangular picture simultaneously provided the edges are gluable.*

The subnet in Fig. 3 illustrates how insertion rule is applied on up, down, left and right simultaneously of a triangular picture q.

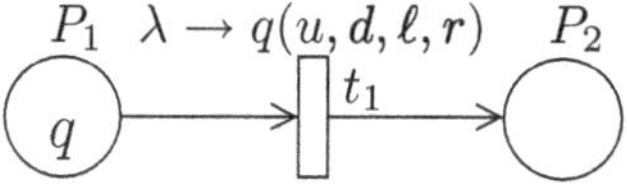

Fig. 3. Subnet used for the following triangular picture representation

The output place will consists of the triangular picture in Fig. 4, where

q = B D C A

Definition 8. *The family of triangular picture languages generated by TATPN with EER and PER on transition is denoted by* $\mathcal{L}(TATPN)$.

Theorem 1. *There exist TATPN with EER and PER on transitions which generate "honey comb pattern" and "swastik pattern" of triangular picture languages.*

Proof. (i) We construct a TATPN which generates honeycomb patterns. Honeycomb materials are widely used where flat or slightly curved surfaces are needed and so they are widely used in the aerospace industry. Honeycomb tiles are used in floor and wall designs.

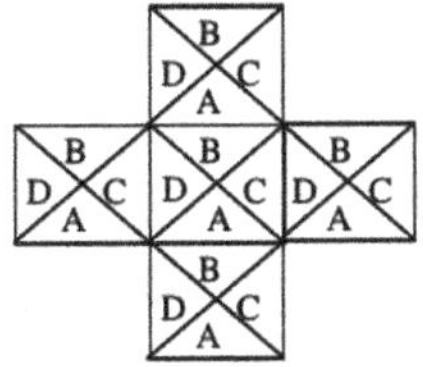

Fig. 4. Triangular picture of subnet in Fig. 3

Let Σ= { A , B }. Using these tiles and Elementary Evolution Rule (EER) and Parallel Evolution Rule (PER) on transitions of a TATPN we generate a honeycomb pattern. Construct the $TATPN_1$ as in Fig. 5.

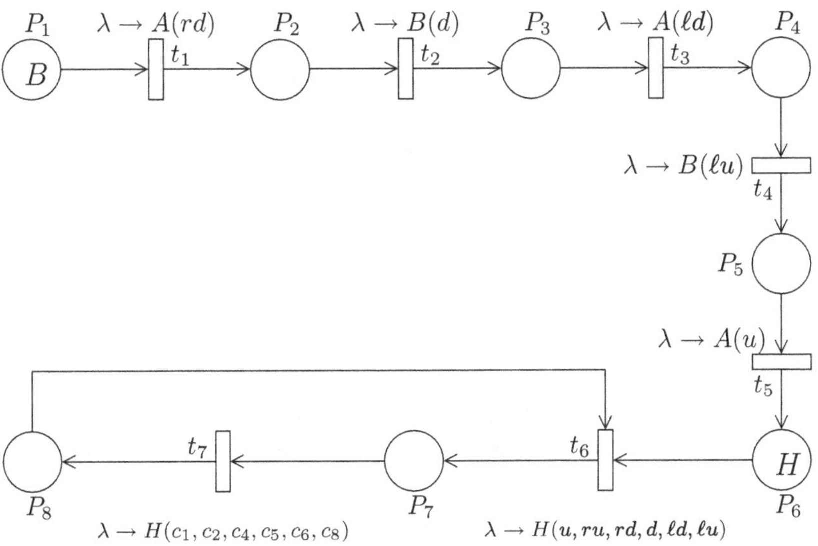

Fig. 5. $TATPN_1$ generating Honeycomb patterns

The derivation is shown in Fig. 6.

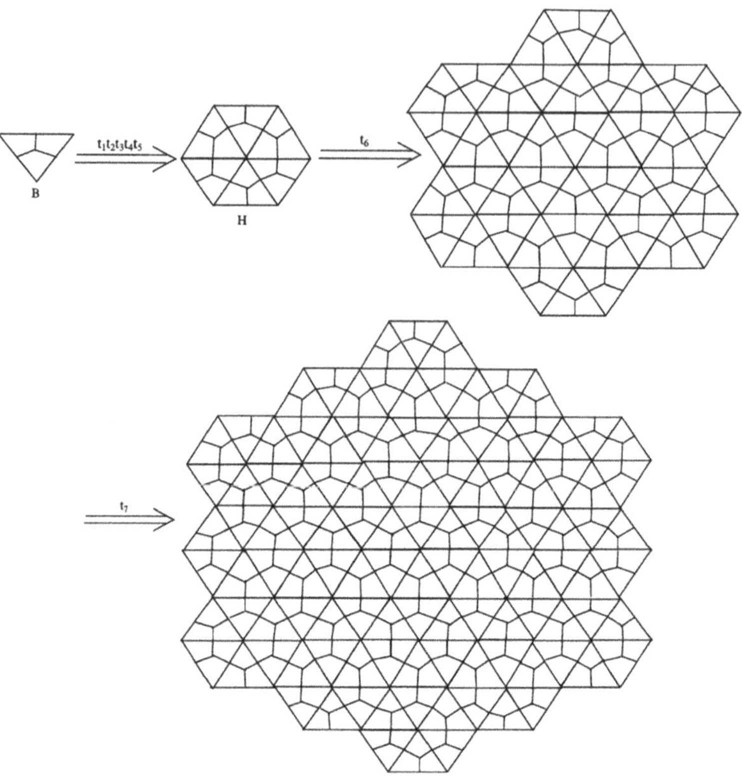

Fig. 6. Honeycomb patterns

We can generate different sizes of this picture as the number of times t_6 and t_7 fire.

(ii) We construct a TATPN which generates Swastik pattern. The swastika is a geometrical figure and an ancient religious icon in the cultures of Eurasia.

Let $\Sigma^T = \{A, B, C, D, E, F, G, H\}$. Using these tiles, EER and PER on transitions of $TATPN_2$ as in Fig. 7, we get the swastik pattern.

In $TATPN_2$, when the transitions $t_1t_2t_3$ fires, we get the pattern in Fig. 8(a) and again when the transition t_4 fires we get Fig. 8(c) and again when the transition t_5 fires, we get Fig. 8(d) and when the transition t_6 fires, we get the swastik pattern in Fig. 8(e) and when the transition t_7 fires, we get the swastik pattern in Fig. 8(f) and when t_8 fires, it removes the token Q from P_8 and it uses the rule $\lambda \to Q(u, r, d, l)$ and deposits the pattern Fig. 9 in P_9.

When t_9 fires we get the swastik pattern as in Fig. 10. We can generate different sizes of this pattern as the number of times t_8 and t_9 fires.

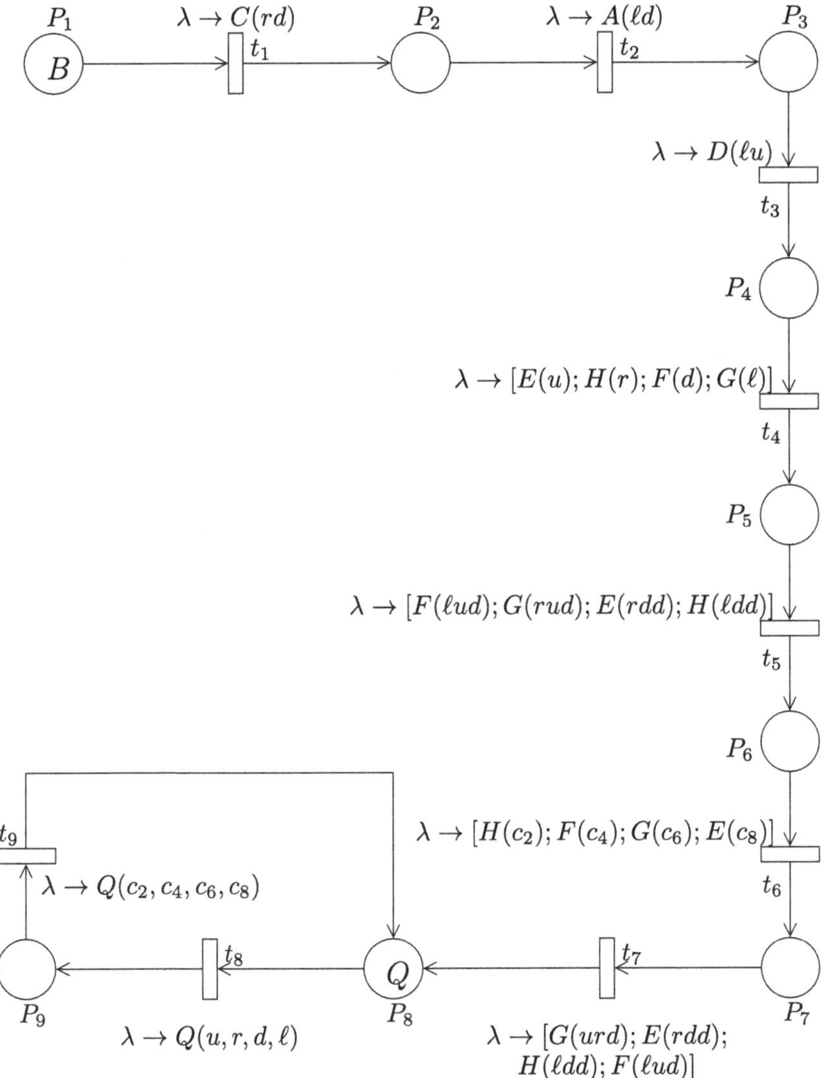

Fig. 7. $TATPN_2$ generating Swastik pattern

4 Triangular Array Token Petri Net *P* System

In this section, we develop a P system called Triangular Array Token Petri Net *P* System (TATPNPS), which uses Elementary Evolution Rules (EER) and Parallel Evolution Rules (PER) introduced in Sect. 3 as evolution rules in its regions and has labeled triangular arrays as objects.

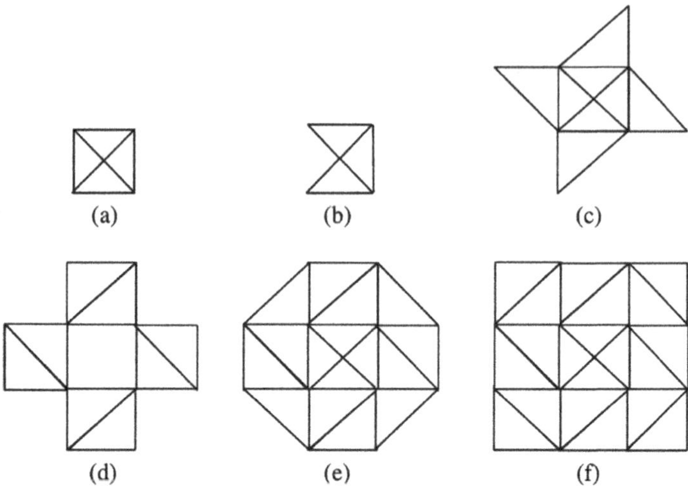

Fig. 8. Patterns which generates Swastik pattern

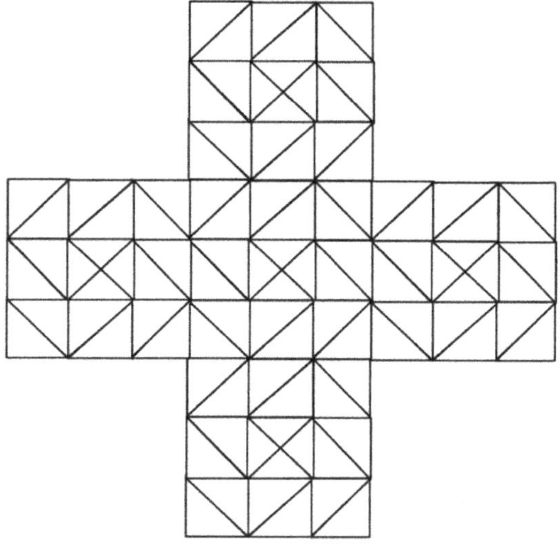

Fig. 9. Pattern which generates Swastik pattern

Definition 9. *A Triangular Array Token Petri Net P System (TATPNPS) is* $\pi = (\Sigma^T, \mu, P_1, P_2, \cdots P_m, (R_1, T_1), (R_2, T_2) \cdots (R_m, T_m), i_0)$ *where* Σ^T *is a finite set of labeled triangular tiles;* μ *is a membrane structure with* m *membranes, labeled in a one- to-one way with* $1, 2, \cdots m$*;* $P_1, P_2, \cdots P_m$ *are finite sets of triangular picture patterns over* Σ^T *associated with the* m *regions of* μ*,* i_0 *is the output membrane which is an elementary membrane and* $R_1, R_2, \cdots R_m$ *are finite sets of evolution rules namely EER and PER associated with the* m *regions of* μ*. The rules* R_i *are of the form* $(r_i,\ tar)$*,* $1 \leq i \leq n$*, where* $r_i \in EER \cup PER$

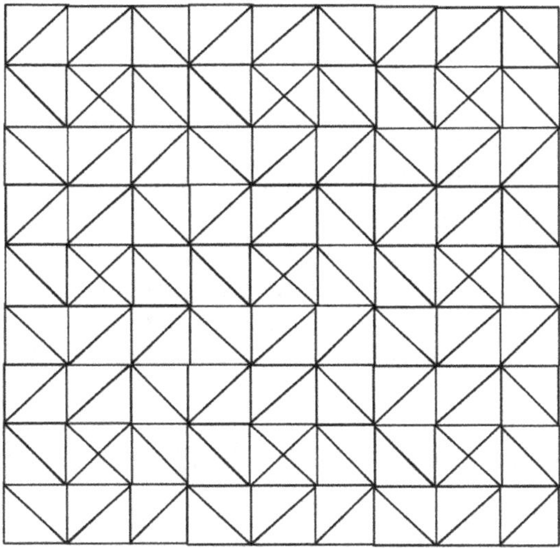

Fig. 10. Swastik pattern using $TATPN_2$

and tar $\in \{in_j, out, here\}$. $T_i, 1 \leq i \leq n$ *are the resultant triangular picture pattern obtained after the evolution rule* R_i *is applied in membrane* i.

The computation process in TATPNPS is defined as to each triangular picture pattern present in the region of the system, the evolution rule associated with respective region should be applied in parallel to the boundary edges of the triangular tiles. Then the resultant triangular picture pattern is moved (remains) to another region (in the same region) with respect to the target indicator in_j *(here) associated with the evolution rule. If the target indicator is out, then the resultant triangular picture pattern is sent immediately to the next outer region of the membrane structure.*

The computation is successful only if the evolution rules of each region is applied. The computation stops if no further application of evolution rule is applicable. The result of a halting computation consists of a triangular picture pattern composed of triangular tiles from Σ^T *placed in the membrane with labeled* i_0 *in the halting configuration. The set of all such triangular picture patterns generated by TATPNPS* π *is denoted by TPL* (π). *The family of all such languages TPL* (π) *generated by system* π *with at most* m *membranes, is denoted by* $TPL_m(TATPNPS)$.

Example 2. Consider the TATPNPS
$\pi_1 = (\Sigma^T, \mu = [_1[_2[_3[_4[_5[_6]_6]_5]_4]_3]_2]_1, P_1, P_2, P_3, P_4, P_5, P_6, (R_1, T_1), (R_2, T_2), (R_3, T_3), (R_4, T_4), (R_5, T_5), (R_6, T_6), 6)$. μ indicates that the system has six regions one within the other, $i_0 = 6$ indicates that the region 6 is the output region.

$\Sigma^T = \{$ A, B, C, D, E, F, G, H $\}$

$$P_1 = \quad , P_2 = P_3 = P_4 = P_5 = P_6 = \phi$$

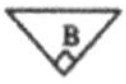

$R_1 = \{(\lambda \rightarrow C(rd), here), (\lambda \rightarrow A(ld), here), (\lambda \rightarrow D(lu), in)\}$
$R_2 = \{(\lambda \rightarrow [E(u); H(r); F(d); G(l)], here),$
$(\lambda \rightarrow [F(lud); G(rud); E(rdd); H(ldd)], in)\}$
$R_3 = \{(\lambda \rightarrow (H(c_2); F(c_4); G(c_6); E(c_8)), in)\}$
$R_4 = \{(\lambda \rightarrow (G(rud); E(rdd); H(ldd); F(lud)), in))\}$
$R_5 = \{(\lambda \rightarrow T_4(u, r, d, l), here), (\lambda \rightarrow T_4(c_2, c_4, c_6, c_8), here),$
$(\lambda \rightarrow T_4(c_2, c_4, c_6, c_8), out), (\lambda \rightarrow T_5(u, r, d, l), here),$
$(\lambda \rightarrow T_5(c_2, c_4, c_6, c_8), here), (\lambda \rightarrow T_5(c_2, c_4, c_6, c_8), out)\}$
$R_6 = \phi.$

The picture derivation is shown in the following table:

Region (i)	Input	Rule (R_i)	Output (T_i)
1	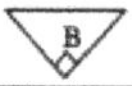	$(\lambda \rightarrow C(rd), here)$	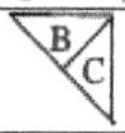
1	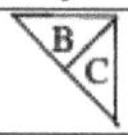	$(\lambda \rightarrow A(ld), here)$	Fig. 8(b)
1	Fig. 8(b)	$(\lambda \rightarrow D(lu), in)$	Fig. 8(a)
2	Fig. 8(a)	$(\lambda \rightarrow E(u); H(r);$ $F(d); G(l), here)$	Fig. 8(c)
2	Fig. 8(c)	$(\lambda \rightarrow F(lud); G(rud);$ $E(rdd); H(ldd), in)$	Fig. 8(d)
3	Fig. 8(d)	$(\lambda \rightarrow H(c_2); F(c_4);$ $G(c_6); E(c_8), in)$	Fig. 8(e)
4	Fig. 8(c)	$(\lambda \rightarrow G(rud); E(rdd);$ $H(ldd); F(lud), in)$	Fig. 8(f)
5	Fig. 8(f)	$(\lambda \rightarrow T_4(u, r, d, l), here)$	Fig. 9
5	Fig. 9	$(\lambda \rightarrow T_4(c_2, c_4, c_6, c_8), here)$	Fig. 10
5	Fig. 10	$(\lambda \rightarrow T_5(u, r, d, l), here)$ $(\lambda \rightarrow T_5(c_2, c_4, c_6, c_8), here)$	Fig. 10 is magnified
		$(\lambda \rightarrow T_4(u, r, d, l), here)$ $(\lambda \rightarrow T_4(c_2, c_4, c_6, c_8), out)$	Fig. 10 is magnified
		$(\lambda \rightarrow T_5(u, r, d, l), here)$ $(\lambda \rightarrow T_5(c_2, c_4, c_6, c_8), out)$	Fig. 10 is magnified
6	Magnified Fig. 10	$R_6 = \phi$	Magnified Fig. 10 swastik pattern

Thus TATPNPS Π_1 generates a family of swastik patterns.

Theorem 2. $TPL_m(TATPNPS) \cap \mathcal{L}(TATPN) \neq \phi$.

Triangular Array Token Petri Net generates Swastik patterns as given in Theorem 1. The same pattern is also generated by Triangular Array Token Petri Net P System in Example 2. Thus the two families intersect.

5 Comparative Study with TTPPS

In this section, TATPNPS and TATPN are compared with Triangular Tile Pasting P System (TTPPS) for generative powers.

Definition 10. *[2] A Triangular Tile Pasting P System (TTPPS) is defined as a five tupule* $\pi = (\Sigma, \mu, F_1, F_2, \ldots F_m, R_1, R_2, \ldots R_m, i_0)$, *where* Σ *is a finite set of iso-triangular tiles,* μ *is a membrane structure. In* μ, *the membranes are labeled in a one-one manner and the labels are* $1, 2, \ldots m$. $F_1, F_2, \ldots F_m$ *are finite sets of pictures over the iso-triangular tiles in* Σ *associated with the* m *regions of the membranes.* $R_1, R_2, \ldots R_m$ *are finite sets of pasting rules of type* $((x_i, y_i), tar)$, $(1 \leq i \leq n)$ *associated with* m *regions of* μ *and* i_0 *is the output membrane, which is an elementary membrane. An evolution in TTPPS is defined in such a way that, to each picture pattern in each region of the system, a pasting rule could be applied and should be applied. The picture pattern is moved (remain) to another region (in the same region) with respect to the target indication associated with the pasting rules.*

A computation is successful only if all the pasting rules are applied. The computation is stopped, if there is no possibility of applying the pasting rules. The result of halting picture pattern is composed only by the pasting rules; the pattern is halted in the output region of membrane, i_0. The set of all picture patterns computed with the pasting rules of TTPPS is denoted by $TTPPL(\pi)$. The set of all languages $TTPPL(\pi)$ generated by the systems π is denoted by $TTPPL_m$.

Example 3. [2] The Triangular Tile Pasting P System

$$\pi_1 = (\Sigma, [_1[_2]_2[_3]_3]_1, F_1, F_2, F_3, R_1, R_2, R_3, 1)$$

generates a class of two dimensional triangular picture language $\mathcal{L}_1$ with atmost 4 distinct labeled iso-triangular tiles.

Here $\Sigma = \{A, A_1, B, B_1\}$, $F_1 = \phi$, $F_2 = A_1$, $F_3 = \phi$.
$R_1 = \{(A, (a_3, b_3), in_2), (B_1, (b_{11}, a_1), here\}$,
$R_2 = \{(B, (b_2, a_2), in_3), (B, (b_1, a_{11}), in_3), (A_1, (a_{13}, b_{13}), in_3)\}$,
$R_3 = \{(B, (b_2, a_2), in_2), (B_1, (b_{12}, a_{12}), out), (A_1, (a_{13}, b_{13}), in_2)\}$,
and 1 is the output region.

The derivation starts in region 2, R_2 is applied and depending on the target indication, the resultant picture pattern moves to another region and finally the resultant iso-triangular picture pattern is collected in the output membrane one.

The first two members of the language $\mathcal{L}_1$ is shown below:

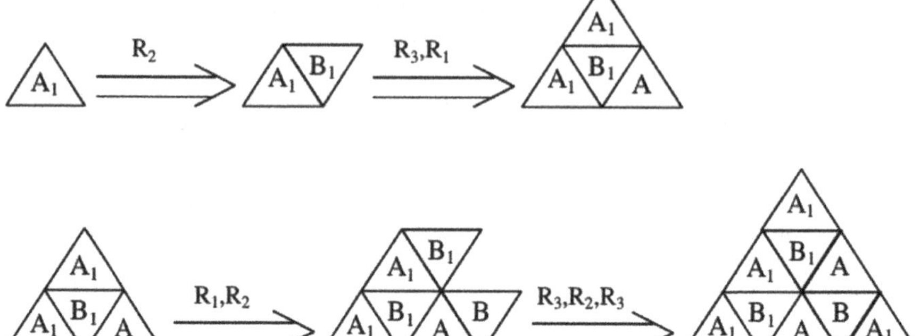

Theorem 3. *The families of languages generated by TATPNPS and TTPPS are incomparable but not disjoint.*

Proof. The families of languages generated by TATPNPS and TTPPS are by parallel mechanism. In TATPNPS the catenation rules namely EER and PER generate the family of languages.

The language of two-dimensional iso-triangular picture pattern $\mathcal{L}_1$ with atmost 4 distinct labeled iso-triangular tiles given in Example 3 cannot be generated by TATPNPS, since different labels cannot be used in catenation rules, i.e. to which tile either A_1 or A, the catenation rule has to be used cannot be distinguished. The catenation rules are mainly along the directions.

The language of swastik patterns given in Example 2, cannot be generated by TTPPS, since EER and PER rules are used. In TTPPS, the tiles are pasted along the gluable edges, whereas in TATPNPS, the tiles are catenated in corners as well as along the directions.

The language of iso-triangular picture pattern with two tiles namely A and B can be generated by both systems.

Consider the TATPNPS $\pi_2 = (\Sigma^T, \mu = [_1[_2]_2]_1, P_1, P_2, (R_1, T_1), (R_2, T_2), 2)$, μ indicates that the system has two regions one within the other, $i_0 = 2$ indicates that the region 2 is the output region.

$$\Sigma^T = \{\text{A}, \text{B}\},\ P_1 = \text{A},\ P_2 = \phi.$$

$$R_1 = \{(\lambda \to B(ru), here), (\lambda \to A(u), here), (\lambda \to A(rd), here), (\lambda \to A(rd), in)\}$$
$$R_2 = \phi.$$

Beginning with the initial object $P_1 =$ A in region 1, the evolution rule R_1 is applied, once the rules $(\lambda \to B(ru), here)$ and $(\lambda \to A(u), here)$ are applied we get a picture pattern as in Fig. 11(a) and when the rule $(\lambda \to A(rd), here)$ is applied we get an iso-triangular picture as in Fig. 11(b) and then the rules in R_1 are applied again and the process continues. When the rule $(\lambda \to A(rd), in)$

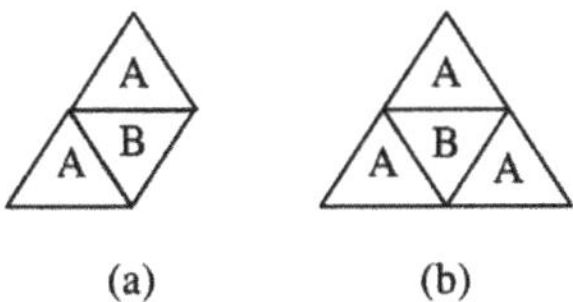

Fig. 11. iso - triangular picture pattern

is applied we get the resultant iso-triangular picture pattern and this pattern reaches region 2, which is the output membrane.

Now, consider the TTPPS $\pi_2 = (\Sigma, [_1[_2]_2]_1, F_1, F_2, R_1, R_2, 2)$ which generates the language of iso-triangular picture pattern.

$\Sigma = \{$ A, B $\}$, $F_1 = A$, $F_2 = \phi$

$R_1 = \{(A, (a_3, b_3), here), (B, (b_2, a_2), here), (B, (b_1, a_1), here), (B, (b_1, a_1), in)\}$
$R_2 = \phi$.

Beginning with the initial object $F_1 =$ A, the rules of R_1 are applied and the iso-triangular picture pattern as given in Fig. 11(a) and 11(b) are generated, which is a first member of the language and the process continues till the rule $(B, (b_1, a_1), in)$ is applied. Once this rule is applied the resultant picture is sent to region 2, which is the output membrane.

Theorem 4. *The families of languages generated by TATPN and TTPPS are incomparable but not disjoint.*

Proof. The language of two dimensional iso-triangular picture pattern $\mathcal{L}_1$ with atmost 4 distinct labeled iso-triangular tiles given in Example 3 cannot be generated by TATPN. The language of swastik patterns given in Example 2, cannot be generated by TTPPS since EER and PER rules are used in TATPN.

The language of iso-triangular picture pattern with two iso-triangular tiles namely A and B can be generated by both systems. TTPPS generating this language is given in Theorem 3. Now, consider the TATPN generating this language.

Let $\Sigma = \{$ A, B $\}$, using these tiles and EER rules on transitions of a TATPN, we get the desired pattern. Construct $TATPN_3$ as in Fig. 12.

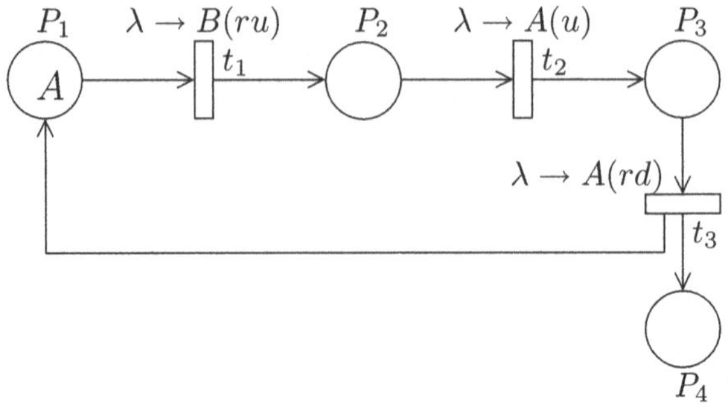

Fig. 12. $TATPN_3$ generating iso-triangular picture pattern

6 Conclusion

In this paper, we have generated honeycomb and swastik patterns of triangular picture languages using some evolution rules on transitions of TATPN. We have also introduced TATPNPS and it is shown that the two families intersect. It is worth examining the superiority between TATPN and TATPNPS with regard to their generative powers. It is our future work. We compared TATPN and TATPNPS with TTPPS for their generative powers. Generation of other interesting Kolam patterns using TATPNPS is another direction of study with applications in image analysis.

References

1. Bhuvaneswari, K., Kalyani, T., Lalitha, D.: Triangular tile pasting P system and array generating Petri nets. Int. J. Pure Appl. Math. **107**(1), 111–128 (2016)
2. Bhuvaneswari, K., Kalyani, T., Thomas, D.G., Lalitha, D.: *P* systems on iso-triangular arrays. Italian J. Pure Appl. Math. **45** (2021)
3. Ceterchi, R., Mutyam, M., Paun, G.: Array-rewriting P systems. Nat. Comput. **2**, 229–249 (2003)
4. Immanuel, B., Rangarajan, K., Subramanian, K.G.: String token petri nets. In: Proceedings of the European Conference on Artificial Intelligence, One Day Workshop on Symbolic Networks, at Vanlencia, Spain (2004)
5. Immanuel, B., Usha, P.: Array-token petri nets and 2D grammars. Int. J. Pure Appl. Math. **101**(5), 651–659 (2015)
6. Kalyani, T., Sasikala, K., Thomas, D.G., Robinson, T., Nagar, A.K., Paramasivan, M.: 3D-array token petri nets generating tetrahedral picture languages. In: Lukić, T., Barneva, R.P., Brimkov, V.E., Čomić, L., Sladoje, N. (eds.) IWCIA 2020. LNCS, vol. 12148, pp. 88–105. Springer, Cham (2020). https://doi.org/10.1007/978-3-030-51002-2_7
7. Kamaraj, T., Lalitha, D., Thomas, D.G., Thamburaj, R., Nagar, A.K.: Adjunct hexagonal array token petri nets and hexagonal picture languages. Math. Appl. **3**, 45–59 (2014)

8. Kannamma, S., Rangarajan, K., Thomas, D.G., David, N.G.: Array Token Petrinets, Computing and Mathematical Modeling, pp. 299–306. Narosa Publishing House, New Delhi (2006)
9. Kuberal, S., Kamaraj, T., Kalyani, T.: Octagonal arrays and petri nets: a computational ecology approach. Ekoloji **28**, 743–751 (2019)
10. Lalitha, D., Rangarajan, K.: Petri nets generating Kolam patterns. Indian J. Comput. Sci. Eng. **3**(1), 68–74 (2012)
11. Lalitha, D., Rangarajan, K., Thomas, D.G.: Rectangular arrays and petri nets. In: Barneva, R.P., Brimkov, V.E., Aggarwal, J.K. (eds.) IWCIA 2012. LNCS, vol. 7655, pp. 166–180. Springer, Heidelberg (2012). https://doi.org/10.1007/978-3-642-34732-0_13
12. Paun, Gh.: Computing with membranes. J. Comput. Syst. Sci. **61**(1), 108–143 (2000)
13. Peterson, J.L.: Petri Net Theory and Modeling of Systems. Prentice Hall Inc., Englewood Cliffs (1981)
14. Rosenfeld, A., Siromoney, R.: Picture languages - a survey. Lang. Des. **1**, 229–245 (1993)
15. Sharon Philomena, V., Usha, P., Santhiya, R.: Generation of certain patterns using array-token petrinets. Int. J. Comput. Sci. Eng. **7**(Special issue 5), 25–29 (2019)
16. Subramanian, K.G., Saravanan, R., Thamburaj, R.: P systems for array generation and application to Kolam patterns. Forma **22**, 47–54 (2006)
17. Subramanian, K.G., Bera, S., Song, B., Pan, L., Zhang, Z.: Array P systems for array based on parallel rewriting with array contextual rules. In: Pre-Proceedings of Asian Conference on Membrane Computing (ACMC 2017), pp. 403–415 (2017)

P System as a Computing Tool for Embedded Feature Selection and Classification Method for Microarray Cancer Data

Ravie Chandren Muniyandi(✉) and Naeimeh Elkhani

Research Center for Cyber Security, Faculty of Information Science and Technology, Universiti Kebangsaan Malaysia, 43600 Bangi, Selangor, Malaysia
ravie@ukm.edu.my, naeimeh.elkhani@siswa.ukm.edu.my

Abstract. Selection of relevant genes is the crucial task for sample classification in microarray data, where researchers try to identify the smallest possible set of genes that can still achieve good predictive performance. Due to the problem of higher risk of overfitting in wrapper methods and sensitivity of the best embedded way to filter out factor that leads to unstable model and significantly different gene subsets, in this paper, we propose a novel model for evaluating and improving techniques for selecting informative genes from microarray data. This model inspired by membrane computing and used the kernel P system (kP) as the variant of the P system to improve the performance of the intelligent algorithm, multi-objective binary particle swarm optimization (MObPSO). The proposed model consists of two main parts. First, kP-MObPSO, which resembles a wrapper type feature selection, and the second part that improves the results of the first part through an embedded feature selection and classification idea based on the kP system. Division, rewriting, and input/output rules are used to make interaction among the genes inside and between the particles. The proposed model applied to the colorectal and breast dataset contains 100 genes with six attributes. The embedded part of the model extracts the marker gene sets indicate more stability and reliability based on ROC measure as well as better error rate in comparison to the wrapper part of the model. In the paper, the lowest error rate by an embedded model is displayed as 0.1111 for breast cancer and 0.0769 for colorectal data.

Keywords: Embedded feature selection · Classification · Microarray cancer data · Kernel P system · Multi-objective binary particle swarm optimization

1 Introduction

Membrane Computing (MC) is a bio-inspired computing model based on the processes taking place in the compartmental structure of a living cell [1]. The devices

R. Freund et al. (Eds.): CMC 2020, LNCS 12687, pp. 94–125, 2021.
https://doi.org/10.1007/978-3-030-77102-7_6

of this model are generically called P Systems. P systems evolve by repeatedly applying rules and mimicking chemical reactions, such as transportation across membranes, cellular division, or death processes, and halt when no more rules can be applied. The essential features of the P system include communication rule, parallelism, and non-determinism. The origins of P systems make it highly suitable as formalism [2] for representing biological systems, especially multicellular systems and molecular interactions taking place in different locations of living cells. For example, membrane computing was applied to improve the efficiency and effectiveness of optimization problems [3], HIV-infection simulations [4], avascular tumor-growth modeling [5], modeling the epidermal growth factor for receptor signaling network [6], quorum-sensing P systems [7], and converting differential-equation models of biological systems to membrane computing [8].

Application of membrane computing to a cancer diagnosis is a new and worthwhile field, given that cancer is one of the leading health issues in many countries all over the world due to the difficulties associated with early diagnosis, which is required to begin the treatment at the initial stage. Although there was a continuous decline in fatality rates related to cancer cases over the last 20 years, from 215.1 deaths per 100,000 people in 1991 to 171.8 in 2010 [9], this was primarily due to screening and early diagnosis rather than a capability to manage and cure the disease effectively. Recently, new models of P systems have been explored. A kernel P system (kP system) based on the tissue P system (graph-based) has been defined. This consists of a low-level specification language that uses established features of existing P-system variants and also includes some new elements. Importantly, kP systems offer a coherent way of integrating these elements into the same formalism [10].

Inspired by Artificial Intelligence algorithms for feature selection of cancer microarray data, here, we propose a membrane-inspired feature-selection method to use the potentials of membrane computing, such as decentralization, non-determinism, and maximal parallel computing, to address the limitations of AI-feature selection. A kP system was used to model the MObPSO model in sequential execution. Particle swarm optimization has been using to develop feature selection methods in microarray gene expression studies [11–13]. Graph-based MObPSO [14] was modeled for the following reasons: 1) its ability to model genes (nodes) and define relationships between them (edges); 2) it has a higher accuracy as compared with flat (filter and wrapper) methods, Sequential Backward Elimination (SBE), Correlation-based Feature Selection (CFS), minimum Redundancy Maximum Relevance (mRMR), and Sequential Forward Search (SFS); and 3) the time complexity of our method on CPUs was reasonable and no higher than other comparable techniques. Then, the SVM classifier was implemented by kP rules to evaluate the classification accuracy of the marker genes resulted from the proposed.

From the perspective that defines how feature selection and classification methods are combined, our proposed kP-MObPSO-SVM computation method belongs to wrapper methods. The advantage of wrapper methods is that they overcome the problem of ignoring feature dependencies that occur in filter methods through several multivariate filter techniques. Filter methods do not incorporate learning. Wrapper methods use machine learning to measure the quality of subsets of features without integrating knowledge about the specific structure of the classification or regression function. They can, therefore, be combined with any machine learning. Still, there is a third type in feature selection and classification methods known as the embedded method. Embedded methods differ from other feature selection methods in the way feature selection, and learning interact. In contrast to filter and wrapper approaches, in embedded methods, the learning part and the feature selection part cannot be separated, the structure of the class of functions under consideration plays a crucial role. Particle swarm optimization has been using in microarray cancer and to build wrapper and filter feature selection methods [15–17].

One of the famous embedded methods in cancer classification is SVM-Recursive Feature Elimination (SVM-RFE) [18]. According to Tang *et al.* [19], SVM-RFE is unstable because it is highly sensitive to f ("filter out" factor), and different f result in significantly different gene subsets, which in turn result in SVM classifiers with significantly different testing accuracies. For example, when the factor is 0.7, 0.3, or -1, the testing accuracy is 85.29% for 64 genes; however, when the factor is 0.8, 0.5, or 0.1, the testing accuracy is 94.12% for 64 genes. A careful exploration of reasons why the SVM-RFE algorithm is sensitive to f, should be helpful to find a more reliable algorithm for gene selection and cancer classification. Here "reliable" means "accurate and stable". Random forest and multiple feature selection and classification are used to develop embedded methods in microarray data [20,21].

In this paper, to improve the performance of the proposed wrapper kP-MObPSO, we added a second part to our proposed model, which is inspired by the embedded method. The marker gene sets resulting from the first part of the proposed model (kP-MObPSO) will use for further evaluation to add the absent genes to the groups of marker genes and consequently evaluate the classification result. The classification uses a performance evaluation measure (accuracy Eq. (1)), but it is not enough to decide for the crucial case, especially in medical diagnosis. Therefore, it is also suggested another type of performance evaluation tools such as the ROC (receiver operation characteristics) and F1-measure. This study presents a comparison of the performances of the wrapper kP-MObPSO and embedded kP-MObPSO using the open-source software WEKA in terms of accuracy, specificity, sensitivity, F-measure, and ROC curve. Accuracy, defined in Equation (1) (TN = True Negative, TP = True Positive, FN = False Negative, FP = False Positive), is used as a metric for performance evaluation.

$$Accuracy = (TN + TP)/(TN + FN + FP + TP) \tag{1}$$

If the newly added gene were able to improve the classification measures like accuracy, error rate, and ROC, the absent gene would keep in the gene set; otherwise, feature selection will continue to test the rest of the genes out of the total number of genes those are absent in the marker gene sets. Within the process of embedded part error rate calculation involves classification, which is based on SVM written by KP system rules.

2 Previous Approaches

The proposed models in this study are based on the preliminary concepts and algorithms in multi-objective binary particle swarm optimization, support vector machine, and kernel p system. MObPSO is already updated and improved through the rules of the kernel p system so-called kP-MObPSO. To execute SVM on the kP-MObPSO, kernel p system rules are used to define SVM on kP-MObPSO. Correlation-based feature selection is used to analyze microarray gene expression data.

2.1 Microarray Gene Expression Data Analysis for Cancer Classification

The genetic information of cells is stored in deoxyribonucleic acid (DNA), and all cells in an organism have the same genome. However, due to different tissue types, different development stages, and even other environmental conditions, genes from cells in the same organism can be expressed in various combinations and/or different quantities during transcription from DNA to messenger ribonucleic acid (mRNA). Other organisms have different genomes and distinct gene expression patterns.

Recently, gene expression microarrays including the complementary DNA microarray (cDNA) and the GeneChip have been developed as a powerful technology for functional genetics studies, which simultaneously measures the mRNA expression levels of thousands to tens of thousands of genes. A typical microarray expression experiment monitors the expression level of each gene multiple times under different conditions or in different phenotypes. For example, a comparison can be made between healthy tissue and cancerous tissue, or one kind of cancerous tissue versus another. By collecting such huge gene expression data sets, it opens the possibility to distinguish phenotypes and to identify disease-related genes whose expression patterns are excellent diagnostic indicators [22].

2.2 Correlation-Based Feature Ranking Algorithms for Gene Selection

A common disadvantage of filter models is that they ignore the interaction with the classifier (the search in the feature subset space is separated from the search in the hypothesis space) and that most proposed techniques are univariate. This means that each feature is considered separately, thereby ignoring feature dependencies, which may lead to worse classification performance when compared to other types of feature selection techniques. To overcome the problem of neglecting feature dependencies, several multivariate filter techniques were introduced, aiming at the incorporation of feature dependencies to some degree. These techniques are elaborated by Hall, Koller *et al.* and Yu *et al.* [23–25].

Gene selection can be viewed as a feature selection or dimensionality reduction problem. Currently, there are mainly two kinds of algorithms for gene selection: correlation-based algorithms and back elimination algorithms. Correlation-based feature ranking algorithms work in a forward selection way by ranking genes individually in terms of a correlation-based metric. Then some top-ranked genes are selected to form the most informative gene subset [26]. One of the common matrices is signal to noise which will be used in our study - Signal-to-Noise (S2N) [17]:

$$\omega_i = |\mu_i(+) - \mu_i(-)|/\sigma_i(+) + \sigma_i(-) \quad (2)$$

In Eq. (2), $\mu_i(+)$ and $\mu_i(-)$ are the mean values of the ith gene's expression data over positive and negative samples in the training dataset, respectively. $\sigma_i(+)$ and $\sigma_i(-)$ are the corresponding standard deviations. Correlation-based algorithms are straightforward and work efficiently. The time complexity is $O(d * \log_2 d)$ for ranking where d is the size of the original gene set. However, a common drawback is that these algorithms implicitly assume that genes are orthogonal to each other and thus can only detect relations between class labels and a single gene.

2.3 The Kernel P System

According to Gheorghe *et al.* [10], A kP system of degree n is a tuple, $k\Pi = (O, \mu, C_1, ..., C_n, i_0)$, where O is a finite set of objects, called an alphabet; μ defines the membrane structure, which is a graph, (V, E), where V represents vertices indicating compartments, and E represents edges; $C_i = (t_i, i), 1 \leq i \leq n$, is a compartment of the system consisting of a compartment type from T and an initial multiset, ω_i, over O; i_0 is the output compartment, where the result is obtained (this will not be used in this study). kP systems use a graph-like structure (similar to that of tissue P systems) and two types of rules:

1. Rules to process objects: these rules are used to transform objects or to move objects inside compartments or between compartments. These types of rules categorize to rewriting, communication, and input-output rules:
   (a) Rewriting and communication rule: $x \rightarrow y\{g\}$, where $x \varepsilon A+$, $y \varepsilon A*$, $g \varepsilon FE(A \cup \bar{A})$; y at the right side defines as $y = (a_1, t_1)...(a_h, t_h)$, where $a_j \varepsilon A$ and $t_j \varepsilon L$, $1 \leq j \leq h$, a_j is an object, and t_j is a target, respectively.
   (b) The input-output rule: $(x/y)\{g\}$, where $x, y \varepsilon A*$, $g \varepsilon FE(A \cup \bar{A})$, means that x can be sent from the current compartment to the environment or y can be brought from the environment to the target compartment.
2. System structure rules: these rules make a fundamental change in the topology of the membranes, for example, with division rule on a compartment, dissolution rule on a specific compartment, make a link between compartments, or dissolve the link between them. These rules are described as follow:
   (a) Division rule: $[\,]_{li} \rightarrow [\,]_{1i_1}, ..., [\,]_{1i_n}\{g\}$, where $g \varepsilon FE(A \cup \bar{A})$; means compartment li can be replaced with n number of compartments. All newly created compartments inherit objects and links of li;
   (b) Dissolution rule:$[\,]_{1i} \rightarrow \lambda\{g\}$ means compartment li does not exist anymore as well as all its links with other compartments.
   (c) Link-creation rule: $[\,]_{1i}; [\,]_{1j} \rightarrow [\,]_{1i}\text{-}[\,]_{1j}\{cg\}$ means a link will be created between compartment li with compartment lj. If there is more than one compartment with the label lj, one of them will have a link with lj non-deterministically.

(d) Link-destruction rule: $[\,]_{1i}$-$[\,]_{1j} \rightarrow [\,]_{1i}; [\,]_{1j} \{cg\}$ means the existence link between li and lj will eliminate and there will not be any link between them anymore. The same as link creation, if there is more than one compartment that has a link with li, then one of them will be selected non-deterministically to apply this rule.

2.4 The MObPSO Approaches

Optimization problems with multiple goals or objectives are referred to as multi-objective optimization (MOO) problems. Therefore, the objectives may estimate different aspects of solutions, which are partially or wholly in conflict. MOO can be defined as follows: optimize $Z = (f_1(x), f_2(x), ..., f_m(x))$, where $x = (x_1, x_2, ..., x_m) \varepsilon X$. A multi-objective searching concept is clearly described by Yu *et al.* [27].

Graph Structure is one of the models of feature selection for classification. For example, a graph-based MObPSO algorithm [14] proposed to optimize the average of node weights and edge weights at the same time through making different subgraphs. This algorithm is a feature selection model to highlight relevant and non-redundant genes in microarray datasets. The results of microarray datasets indicated that graph-based MObPSO produces better performance compared to SBE, CFS, mRMR, and SFS methods from a classification accuracy point of view. Although execution time in such a technique, which is known as optimization techniques, is not efficient, their time complexity is not much higher relative to other comparable methods [14]. MObPSO is designed for maximizing the dissimilarity (negative correlation) and signal-to-noise ratio (SNR), Eqs. (3) (cov = co-variance, var= variance) and (4) (sd = standard deviation), respectively, which are represented as edge weight and node weight, respectively. Arbitrarily selected features initialize the population from the data matrix, and population-fitness values are calculated using dissimilarity and SNR average values. The archive, A, is initialized by the population value after nondominated sorting of the primary population. Velocity and position are updated using Eqs. (5) and (6). If we consider a D-dimensional space, the exact place of each particle will be shown by $x_i = (x_{i1}, x_{i2}, ..., x_{iD})$. In the same way, for each particle their velocity can be written as a vector like $v_i = (v_{i1}, v_{i2}, ..., v_{iD})$. The limitation for the position and velocity of particles represent as $[X_{min}, X_{max}]_D and [V_{min}, V_{max}]_D$, respectively. The value called *pbest* will present the best position which have met by the ith particle previously $p_i = (p_{i1}, p_{i2}, ..., p_{iD})$. The ultimate best value from all of the positions have been met so far will represent by gbest value as $g = (g_1, g_2, ..., g_D)$. In every iteration, the *pbest* and *gbest* value for position and velocity of ith particle will update in the swarm. Where w represents inertia weight to control the effect of particle previous and new velocity, the variables $r1$ and $r2$ represent any random numbers between (0, 1), and constant numbers as $c1$ and $c2$ represent the acceleration. It means these constant numbers will control the distance that a particle can travel far in every iteration. The variables $v(t+1)$ and $v(t)$ indicate the velocities of the new and old particle, respectively.

The local best P is updated after comparing the current and previous fitness values of a particle, and the global best G is updated according to randomly picking a particle from the archive. These steps are repeated for a particular number of iterations. A summary of the MObPSO algorithm is explained in Table 1.

$$dissimilarity = \left(1 - \frac{cov(x,y)}{\sqrt{var(x)var(y)}}\right) \tag{3}$$

$$SNR = \left|\frac{mean(C1) - mean(C2)}{sd(C1) - sd(C2)}\right| \tag{4}$$

$$v(t+1) = w * v(t) + C1 * r1 * (pbest(t) - x(t)) + C2 * r2 * (gbest(t) - x(t)) \tag{5}$$

$$x(t+1) = x(t) + v(t+1) \tag{6}$$

2.5 Embedded Method

Embedded methods differ from other feature selection methods in the way feature selection, and learning interact. Filter methods do not incorporate learning. Wrapper methods use a learning machine to measure the quality of subsets of features without integrating knowledge about the specific structure of the classification or regression function. They can, therefore, be combined with any learning machine. In contrast to filter and wrapper approaches, in embedded methods, the learning part and the feature selection part cannot be separated, the structure of the class of functions under consideration plays a crucial role [28].

Feature selection can be understood as finding the feature subset of a specific size that leads to the largest possible generalization or, equivalently to minimal risk. A vector $\sigma\varepsilon\{0,1\}^n$ model every subset of features of indicator variables, $\sigma_i = 1$ indicating that a feature is present in a subset and $\sigma_i = 0$ indicating that feature is absent ($i = 1, ..., n$). The function G measures the performance of a trained classifier $f^*(\sigma)$ for a given σ. It is vital to understand that although we write $G(f^*, ., ., .)$ to denote that G depends on the classifying or regression function f^*, the function G does not depend on the structure of f^*; G can only access f^* as a black box, for example in a cross-validation scheme. Moreover, G does not depend on the specific learner $\tilde{T}$. In other words $\tilde{T}$, could be any off-the-shelf classification algorithm, and G guides the search through the space of feature subsets. If we allow G to depend on the learner $\tilde{T}$ and on parameters of f^* we get the Eq. (7), as:

$$\overset{inf}{\sigma\varepsilon\{0,1\}^n}\, G(\alpha^*, \tilde{T}, \sigma, X, Y)\; s.t. \begin{cases} s(\sigma) \leq \sigma_0 \\ \alpha^* = \tilde{T}(\sigma, X, Y) \end{cases} \tag{7}$$

Some embedded methods do not make use of a model selection criterion to evaluate a specific subset of features. Instead, they directly use the learner $\tilde{T}$.

Assuming that many learning methods $\tilde{T}$ can be formulated as an optimization problem, we will have Eq. (8) as:

$$\alpha^* = \overset{argmin}{\alpha \varepsilon \Lambda} \; T(\alpha, \sigma, X, Y) = \tilde{T}(\alpha, X, Y) \tag{8}$$

We can rewrite the minimization problem for the particular case of $G = T$ as Eq. (9):

$$\overset{inf}{\alpha \varepsilon \Lambda, \sigma \varepsilon \{0,1\}^n} \; T(\alpha, \sigma, X, Y) \, s.t. \, s(\sigma) \leq \sigma_0 \tag{9}$$

Unfortunately, both minimization Eqs. (8) and (9) are hard to solve. They are existing embedded methods approximate solutions of the minimization problem. One of the ways that embedded methods solve the problem according to Eq. (8) or Eq. (9) is the methods that iteratively add or remove features from the data to greedily approximate a solution of minimization Eqs. (8) or (9). These methods are known as Forward-Backward Methods and can be grouped into three categories. The first category is Forward selection methods, which these methods start with one or a few features selected according to a method specific selection criterion. More features are iteratively added until a stopping criterion is met. The second category is Backward elimination methods that this type starts with all features and iteratively remove one feature or bunches of features. The third category is Nested methods that, during an iteration, features can be added as well as removed from the data.

2.6 SVM for Cancer Classification

SVM is a classification algorithm based on statistical learning theory [29]. Due to extreme sparseness of microarray gene expression data, the dimension of the input space is already high enough so that the cancer classification is already as simple as a linear separable task [30]. It is unnecessary and even useless to transfer it to a higher implicit feature space with a non-linear kernel. As a result, in this work, we adopt linear SVM [31] as the primary classifier as Eq. (10).

$$Linear kernel K(x, y) = <x, y> \tag{10}$$

Where x and y are points in a d-dimensional Euclidean space. For a linear kernel SVM, the margin width can be calculated by Eqs. (11) and (12)

$$w = \sum_{i=1}^{N_s} \alpha_i y_i x_i \tag{11}$$

$$marginwidth = \frac{2}{\|w\|} \tag{12}$$

Where N_s is the number of support vectors, which are defined as the training samples with $0 < \alpha_i \leq C$. SVM is believed to be a superior model for sparse classification problems compared to other models [29,30]. However, the sparseness of a microarray dataset is so extreme that even an SVM classifier is unable to achieve satisfactory performance. A preprocessing step of gene selection is necessary for more reliable cancer classification.

3 Materials and Methods

The proposed model consists of two main parts. Both parts have feature selection and classification components which are implemented through KP system rules. In the first part, kP-MObPSO feature selection finds out the features, and then an SVM based classification, which is implemented by KP rules, will be applied to classify the features. In the second part, feature selection is embedded with classification in a way that, after adding each absent gene to the set of marker genes, came from the first part, the classification will apply immediately to evaluate the difference that can make by the newly added gene. If the newly added gene was able to improve the result of the error rate and ROC, then it will be part of the solution. Otherwise, if it made the result worse, it should be eliminated and embedded feature selection and classification in the second part implemented by kP system rules. Classification also follows SVM through kP rules to evaluate error rate and ROC.

3.1 Description of the Entire Model

The entire model includes two main parts, the first part, features selection based on kP-MObPSO, and then classification based on KP-SVM and the second part, KP-embedded feature selection/SVM Classification. The first part defines modeling and implementing previous MObPSO based on kP system rules with some improvements, which lead to the result consists of a different set of marker genes, so-called kP-MObPSO feature selection. After that, an error rate calculator based on kP-SVM applied to measure the error rate of marker gene sets.

The latter part embedded feature selection of genes with a classification that is based on the kP system and SVM classification. This part starts with the different sets of marker genes that are already highlighted through the first part, and then a mix of feature selection and SVM classification through KP rules on marker genes leads to another set of marker genes, which is called marker genes 2 sets. For doing this, we have followed the third category of embedded methods, which are Nested methods that during an iteration gene can be added as well as removed from the data. Ultimately, new sets of marker genes are supposed to present better error rates in comparison to error rates of marker genes resulted from the first part of the model (features selection based on kP-MObPSO). These two parts are explained in Fig. 1 as the flow chart of the proposed model.

3.2 First Part: Feature Selection Based on kP-MObPSO

The entire process of the proposed kP-MObPSO model is summarized in Table 1. It consists of three main phases, including (i) initialization, (ii) division, (iii) select the minimum $gBestScore$, and (iv) return back the process to the division part again. Each phase is built based on defined objects and rules. Objects are defined as: P: number of particles, Max_c: maximum number of genes inside particles, $position_1, ..., position_n$: n number of positions inside each particle, Reserve, a :

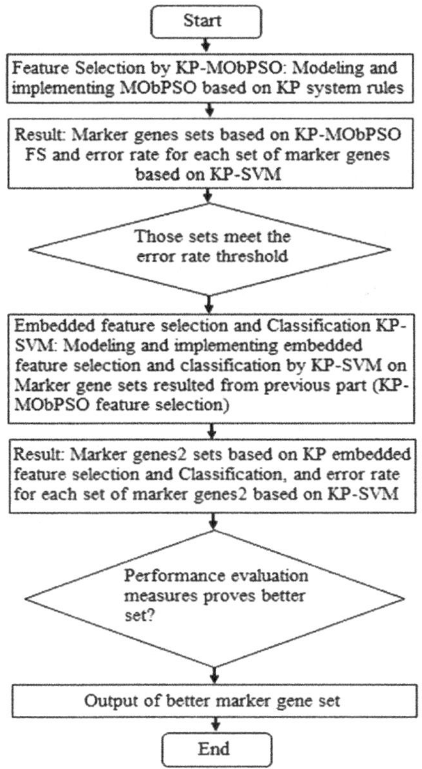

Fig. 1. The flowchart of the proposed model.

$a_1, ..., a_100$: data source of all genes, Max_c:maximum number of genes inside particles, $NGENES, NewNGENES, Q, c, C, sum_{diss}, sum_{snr}, FIT, Q, pBestScore$. Q is a selected gene IDs. Rules are defines based on Table 2.

Table 1. The entire process of the proposed kP-MObPSO model.

Begin
$t = 1$
(i) Initialize
Run all the rules from r_1 to r_{1_6} once
$it = 1$
(ii) Evolution
Run the rules $r_2 > r_3 > r_4 > r_5 > r_6 > r_7 > r_{1_0} > r_{1_1} > r_{1_2} > r_{1_4} > r_{1_5} > r_{1_6}$ in order
till $it = 100$
(iii) $gBestScore = min(Converge)$
(iv) go the section (ii)
till $t = 10$
End

Table 2. Rules kP-MObPSO.

Rules
r_1: Rewriting $r_1 \equiv [[p, max_c]_{position}]_0 \xrightarrow{pos}$ $[[(position^1, ..., position^n)_1, ..., (position^1, ..., position^n)_{position}]_{position}]_0[[\,]_1[\,]_{position}]_0$
r_2: Communication $r_2 \equiv [[(position^1, ..., position^n)_1, ..., (position^1, ..., position^n)_{position}]_{position}]_0 \longrightarrow$ $[[(position^1, ..., position^n)_1, ..., (position^1, ..., position^n)_{position}]_1]_0$
r_3: Communication $r_3 \equiv [(position^1, ..., position^n)_1, ..., (position^1, ..., position^n)_p]_1 \longrightarrow$ $[[(position^1, ..., position^n)_1]p_1, ..., [(position^1, ..., position^n)_p]p_n]_1$ Rules inside each $p : [\,]P_1, ..., [\,]P_n : r_4 > r_5 > r_6 > r_7$
r_4: Rewriting $r_4 \equiv [position, a, max_c, p]^{p_1}_{p_n} \xrightarrow{subgraph1} [NGENES]^{p_1}_{p_n}, [NGENES]^{p_1}_{p_n} \xrightarrow{subgraph1} [NewNGENES, Q, c]^{p_1}_{p_n}$
r_5: Communication/rewriting $r_5 \equiv [NewNGENES, c, p, a]^{p_1}_{p_n} \xrightarrow{MyCost} [C]^{p_1}_{p_n}, [C]^{p_1}_{p_n} \xrightarrow{MyCost} [sum_{diss}]^{p_1}_{p_n},$ $[a]^{p_1}_{p_n} \xrightarrow{MyCost} [snr]^{p_1}_{p_n}, [snr]^{p_1}_{p_n} \xrightarrow{MyCost} [sum_{snr}]^{p_1}_{p_n}, [sum_{diss}, sum_{snr}]^{p_1}_{p_n} \xrightarrow{MyCost} [FIT]^{p_1}_{p_n}$
r_6: Link creation $r_6 \equiv [\,]^{p_1}_{p_n}$ ---$[\,]_{master}$
r_7: Communication/rewriting $r_7 \equiv [FIT^n]^{p_1}_{p_n} \longrightarrow [pBestScore^n]_{master}, [Q^n]^{p_1}_{p_n} \longrightarrow [Q^n]_{master}$
r_8: Division $r_8 \equiv [[[\,]P_1...[\,]P_n[pBestScore_n, Q_n]_{master}]_1]_0 \longrightarrow [[[\,]P_1 ... [\,]P_n[pBestScore_n, gBestScore]_{master}]_{11}[[]P1...[\,]P_n$ $[fitness, pBest, gBest, Velocity, c_1, c_2, w, Vmax, s]_{master}]_{12}]_0$
r_9: Membrane dissolution $r_9 \equiv [[[\,]P_1 ... []P_n[\,]_{master}]_1]_0 \longrightarrow \lambda$
r_{10}: Link creation $r_{10} \equiv [[[\,]P_1 ... [\,]P_n[pBestScore_n]_{master}]_{11}]_0, [[[\,]P_1 ... [\,]P_n[fitness_n]_{master}]_{12}]_0 \longrightarrow$ $[[[\,]P_1 ... [\,]P_n[pBestScore_n]master]_{11}]_0$---$[[[\,]P_1 ... [\,]P_n[fitness_n]_{master}]_{12}]_0$
r_{11}: Communication/rewriting $r_{11} \equiv [[[pBestScore_n]_{master}]_{11}]_0 \longrightarrow [[fitness_n]_{master}]_{11}]_0, [[[pBest_n]_{master}]_{12}]_0 \longrightarrow 1$ $\{[[[fitness_n]_{master}]_{12}]_0 < [[[pBestScore_n]_{master}]_{11}]_0, 1 \le n \le p\}$ & $[[[gBestScore]_{master}]_{11}]_0 \longrightarrow [[[pBestScore_n]_{master}]_{11}]_0, [[[gBest_n]_{master}]_{12}]_0 \longrightarrow 1$ $\{[[[pBestScore_n]_{master}]_{11}]_0 < [[[gBestScore]_{master}]_{11}]_0, 1 \le n \le p\}$ & $[[[converge]_{master}]_{11}]_0 \longrightarrow min[[[pBestScore]^{p_1}_{p_n}]_{master}]_{11}]_0$
r_{12}: Communication $r_{12} \equiv [[[position]^{p_1}_{p_n}]_{12}]_0 \longrightarrow [[[position]^{p_1}_{p_n}]_{master}]_{12}]_0$
r_{13}: Communication/rewriting $[[[position^n, c_1, c_2, c, w, pBest, gBest, p, max_c, rand]_{master}]_{12}]_0 \longrightarrow [[Velocity]_{master}]_{12}]_0$ $[[[Velocity]_{master}]_{12}]_0 \longrightarrow [Vmax]_{master}]_{12}\ \{Velocity > Vmax\}$ $[[[Velocity]_{master}]_{12}]_0 \longrightarrow [-Vmax]_{master}]_{12}\ \{Velocity < Vmax\}$ $[[[[position]^{p_1}_{p_n}]_{master}]_{12}]_0 \longrightarrow 1\ \{rand \le 1/(1 + exp(-2 * Velocity))\}$ $[[[[position]^{p_1}_{p_n}]_{master}]_{12}]_0 \longrightarrow 0\ \{rand > 1/(1 + exp(-2 * Velocity))\}$

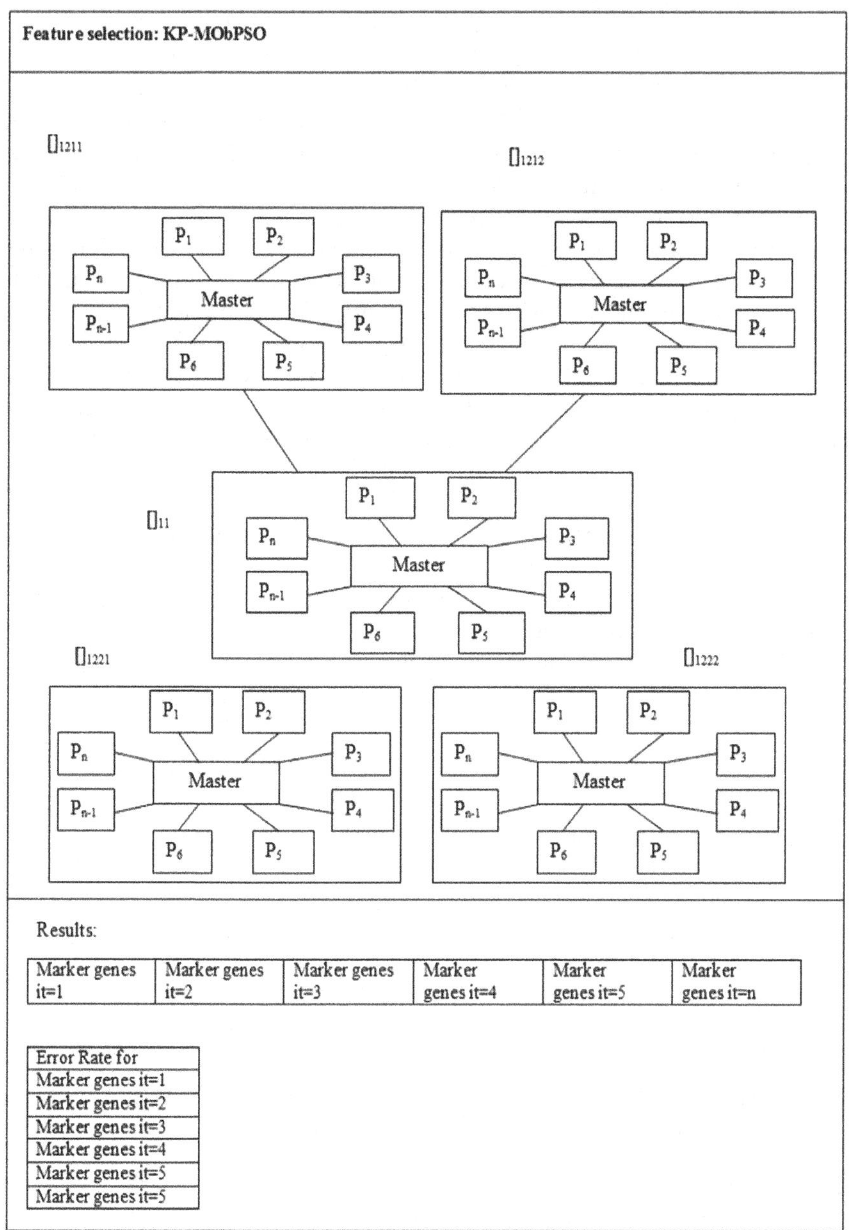

Fig. 2. The results of feature selection by kP-MObPSO.

The results of feature selection by kP-MObPSO are shown in Fig. 2. The division process will take steps as follow: the first compartment 1 will divide into two compartments, namely compartment 11 and 12. Compartment 12 will divide into compartment 121 and 122. After that, compartment 121 will divide into compartment 1211 and 1212, followed by dividing compartment 122 to compartments 1221 and 1222. In between dividing rules, the dissolving rule will apply on compartment 1, 12, 121, and 122, respectively. Thus, as the drawing of the compartment below shows it, compartments 11, 1211, 1212, 1221, and 1222 will remain. The creation of compartments as a result of division rule and dissolution of compartments, which does not have a further contribution in the computation, will execute in parallel whit object processing rules. Object processing rules are rewriting, communication, and input/output rules that make a comparison between assumed gbestscore value and new pbestscore value. This comparison is ultimately between compartment 11 and compartments 1211, 1212, 1221, and 1222 respectively, to replace the better gbestscore in compartment 11. In each step of comparison, for example, between compartment 11 and 1211, the minimum gbestscore represents a set of marker genes. These sets of marker genes produce the final results, as is expected from the kP-MObPSO feature selection model. Sets of marker genes are shown as $(markergenes, it = 1/markergenes, it = 2/.../markergenes, it = n)$ in the figure drawn below. Classification error rate assigned to each set of marker genes. A kP-system based classification consists of the SVM package in Matlab will evaluate the error rate of the marker gene. The rules and objectives of kP-classification will explain in the following.

r_{14}:Output/link
$[[[[position]^{p_1}_{p_n}]_{master}]_{12}]_0 \longrightarrow [[[position]^{p_1}_{p_n}]_{12}]_0 \quad [[]_{12}]_0, [[]_{position}]_0 \longrightarrow []_{12}]_0 \text{- - -} [[\,]_{position}]_0$
$[[[position]^{p_1}_{p_n}]_{1}2]_0 \longrightarrow [[position]^{p_1}_{p_n}]_{position}]_0$
r_{15}: Division rule
$[[\,]p_1 \cdots [\,]p_n[]_{master}]_{12}[[]p_1 \cdots [\,]p_n[]_{master}]_{121}[[\,]p_1 \cdots []p_n[]_{master}]_{122}$
r_{16}: Membrane dissolution
$[[\,]p_1 \cdots [\,]p_n[]_{master}]_{12} \longrightarrow \lambda$

3.3 kP Rules to Find Out Marker Genes and kP-SVM Rules to Calculate Error Rates

A kP-system based computation rules are proposed to find out marker genes and kP-SVM rules used to calculate error rates of marker genes which are resulted from kP-MObPSO. The objects and rules are defined in Tables 3 and 4.

Table 3. The objects to find out marker genes and to calculate error rates of marker genes.

Objects	Details
$(a(i,j)$	Dataset of real data, Index of genes from 1 to 100, $1 \leq i \leq 100$ and j is number of samples $1 \leq j \leq 6$
$Markergenes = zeros(100, n)$	Matrix marker genes will keep the index of marker genes from 1 to 100, $1 \leq i \leq 100$, n = number of particles
$y = zeros(100, 1)$	Matrix will give the flag to marker genes; for gene indexes, $i \leq 50$ flag will be $+1$, and for gene indexes $i > 50$ flag will be -1. It means for normal genes $+1$ and tumor genes -1. In our data set, 50 first genes are normal, and the second 50 genes are tumor genes
$Badgenes = zeros(250, 1)$	Matrix keeps the number of tumor genes is selected for further process
max_j, max_f	Counter for the number of flags, which are $+1$ and -1 by max_j and max_f, respectively
$Y = zeros((max_j + max_f) * 3, 1)$	Dataset consists of 3 samples of data; object Y distributes the flags $+1$ and -1 for the indexes of normal and cancerous genes in the dataset. $+1$ will be distributed max_j times, and -1 will distribute max_f times
$Wholedata = zeros((max_j + max_f) * 3, 1)$	Object whole data is a matrix of real data consists of 3 samples datasets for those genes that are selected as marker genes by index i
$ERR = zeros(150, 2)$	Keeps error rates
$K = 1$	Counter of real data when data adds to $Wholedata$ matrix or deduct from $Wholedata$
M	Matrix of genes

According to Table 4, rule number one, $m1$, calculates a histogram of marker genes and keeps the number of genes repetition in the $table(i)$ for each gene with the index i. m2, rewrites the matrix of marker genes with those genes that are already repeated more than the indication of iteration that is already executing. It means if the code is running on the second iteration, the number of repletion for gene index should be more than two to be considered as a marker gene and so on. The real dataset is built by 50 normal genes value in 6 samples at the beginning followed by 50 cancerous genes value after. $m3$ and $m4$, give a flag to index of genes based on the type of genes, whether they belong to normal genes or cancerous genes as +1 and −1, respectively. The object max_j and max_f updates the number of normal and cancerous genes. $m5$, is the rewriting rule to keep the number of cancerous genes that are already contributing to evaluation. $m6$ reserve a place for the samples of gene indexes are selected as marker genes and $m7$, inputs the real value of reserved samples inside a compartment called whole data. The whole data compartment keeps real data samples for gene indexes that are already highlighted as marker genes. $m8$, applies an SVM package in MATLAB with rewriting rules to evaluate error rates. $m9$, keeps each set of marker genes inside a compartment. Finally, $m10$, indicates how real dataset is divided to n number of compartments called $(particle2, it = 1/particle2, it = 2/.../particle2, it = n)$.

Table 4. The rules to find out marker genes and to calculate error rates of marker genes.

No.	Feature selection (FS): KP rules to get marker genes and calculate error rates (priority based on the rules No.)	KP rules name
m1	$s \to find(M(:) == i), table(i) \to lenght(s), \{1 \leq i \leq 100\}$	Rewriting
m2	$markergenes(i, it+1) \to i\{table(i) > it, 1 \leq i \leq 100, 1 \leq it \leq n\}$	Rewriting
m3	$y(i,1) \to +1 \,\&\, max_j \to max_j + 1, \{1 \leq i \leq 50\}$	Rewriting
m4	$y(i,1) \to -1 \,\&\, max_f \to max_f + 1, \{51 \leq i \leq 100\}$	Rewriting
m5	$BadGenes(it+1, 1) \to max_f$	Rewriting
m6	$Y(j,1) \to +1, \{1 \leq j \leq max_j * 3\}\ Y(j,1) \to -1, \{max_j * 3 + 1 \leq j \leq max_j * 3 + max_f * 3\}$	Rewriting
m7	$wholedata(k,1) \to a(i,j), \{1 \leq i \leq 100, table(i) > it, j = 1\}\ k \to k+1, wholedata(k,1) \to a(i,j), \{1 \leq i \leq 100, table(i) > it, j = 3\, or\, 5\}\ wholedata(k,2) \to a(i,j), \{1 \leq i \leq 100, table(i) > it, j = 2\, or\, 4\}\ wholedata(k,2) \to a(i,j), k \to k+1\{1 \leq i \leq 100, table(i) > it, j = 6\}$	Input
m8	$\{wholedata(k,1), wholedata(k,2)\} \to X\{Y, Holdout = 0.10\} \to P\{cvpartition\}\{X(P.training), Y(P.training)\} \to SVMStruct\{svmtrain\}\{SVMStruct, X(P.test)\} \to C\{svmclassify\}Sum(Y(P.test) \cong C)/P.testsize \to errRateY(P.test), C \to conMatERR(1,1) \to errRate\{first\, 100\, times\, iteration, gBestScore = inf\}ERR(it+1,1) \to errRate \,\&\, Constant \to ERR(it+1,1)\{not\, first\, 100\, times\, iteration, gBestScore \neq inf\}$	Rewriting
m9	$Particle2 \to Markergenes(:, it+1)\{1 \leq it \leq n, 0 < ERR(it+1,1) \leq 0.3\}$	Rewriting
m10	$a(i,j) \to \{(Particle2, it = 1), (Particle2, it = 2), ..., (Particle2, it = n)\}$	Division

3.4 Second Part: Embedded Feature Selection and Classification kP Rules

According to the rules in Table 5, $f1$ is a flag which is an object with a default value zero. $f2$, flag value will rewrite to 1 if it meets the guard values, including an error rate between 0 and 0.3, as well as having at least two cancerous genes indexes in marker genes. $f3$, q, and e are the counters of normal genes and cancerous genes, respectively, which rewrite to a default value zero, and marker genes 2 keep a backup of marker genes resulted from the first part of the model. Marker genes 2 objects will be used for further procedure. For each gene number from 1 to 100, rules number $f4$ to $f10$ will apply to see whether entering a new gene can improve the error rate of that particle or no. $f4$ and $r5$ give a flag to index of genes based on the type of genes, whether they belong to normal genes or cancerous genes as $+1$ and -1, respectively. In parallel, q and e, which are the counters of normal genes and cancerous genes, will be updated. $f6$, the object max_j, and max_f updates the number of normal and cancerous genes. $f7$, clear the value of q and e. $f8$ reserve a place for the samples of gene indexes are selected as marker genes and $f9$, inputs the real value of reserved samples inside a compartment called *wholedata*.

Wholedata compartment keeps actual data samples for gene indexes that are already highlighted as marker genes. $f10$, applies an SVM package in Matlab with rewriting rules to evaluate error rates. To decide whether adding a new gene can improve the error rate or no, rule number $f11$ compares error rates after adding each gene (from 1 to 100) with the constant error rate object resulted in the first part of modeling for each set of marker genes. If adding a new can improve the continuous error rate of that set of marker genes, the $f11$ will add this gene index to the set of marker genes. Otherwise, if a gene value already enters the particle cannot improve the error rate, it should be the exit from the particle. Rules, $f12$ to $f15$ apply output and rewriting rules to eliminate gene values inside the particle. After executing the computation model for n number of iterations which each iteration represents one particle from $it = 1$ to $it = n$, particles of marker genes will update by new genes and result in a new set of particles as $(particle3, it = 1/particle3, it = 2/.../particle3, it = n)$. It means by the end of embedded feature selection and classification, the real data set will divide into a new compartment.

4 Results

The process of calculating error rates from marker genes is explained in Fig. 3, which is a real data example from the colorectal dataset. The method defines 5 sets of marker genes as it=1:5. To get marker genes of these 5 sets, kP-MObPSO will execute 100 times for each set. After extracting marker genes for each set, the error rate will be calculated to indicate which set of marker genes provides a better error rate. An example of real data from colorectal datasets and the procedure of embedded feature selection and classification is explained in Fig. 4.

Table 5. Embedded feature selection and classification KP rules.

No.	Embedded feature selection and classification KP rules to get marker genes 2 and error rates (priority based on the rules No.)	KP rules name
f1	$flag \rightarrow 0$	Rewriting
f2	$flag \rightarrow 1\{0 < ERR(it+1,1) \leq 0.3 \,\&\, BadGenes(it+1,1) \leq 2, 1 \leq it \leq n\}$	Rewriting
f3	$q \rightarrow 0 \,\&\, e \rightarrow 0 \,\&\, markergenes2(:, it+1) \rightarrow markergenes(:, it+1)\{1 \leq it \leq n, 1 \leq i \leq 100\}$	Rewriting
f4	$y(i,1) \rightarrow +1 \,\&\, q \rightarrow q+1\{1 \leq i \leq 50, 1 \leq it \leq n, markergenes(i, it+1) = 0\}$	Rewriting
f5	$y(i,1) \rightarrow +1 \,\&\, e \rightarrow e+1\{51 \leq i \leq 100, 1 \leq it \leq n, markergenes(i, it+1) = 0\}$	Rewriting
f6	$max_j \rightarrow max_j + q, \; max_f \rightarrow max_f + e$	Rewriting
f7	$q \rightarrow 0 \,\&\, e \rightarrow 0$	Rewriting
f8	$Y(j,1) \rightarrow +1\{1 \leq j \leq max_j * 3\} Y(j,1) \rightarrow -1\{max_j * 3 + 1 \leq j \leq max_j * 3 + max_f * 3\}$	Rewriting
f9	$wholedata(k,1) \rightarrow a(i,j)\{1 \leq i \leq 100, j = 1\}\, k \rightarrow k+1, wholedata(k,1) \rightarrow a(i,j)\{1 \leq i \leq 100, j = 3 \, or \, 5\}\, wholedata(k,2) \rightarrow a(i,j)\{1 \leq i \leq 100, j = 2 \, or \, 4\}\, wholedata(k,2) \rightarrow a(i,j), k \rightarrow k+1\{1 \leq i \leq 100, j = 6\}$	Input
f10	$\{wholedata(k,1), wholedata(k,2)\} \rightarrow X\{Y, Holdout = 0.10\} \rightarrow P\{cvpartition\}\{X(P.training), Y(P.training)\} \rightarrow SVMStruct\{svmtrain\}\{SVMStruct, X(P.test)\} \rightarrow C\{svmclassify\} Sum(Y(P.test) \cong C)/P.testsize \rightarrow errRate Y(P.test), C \rightarrow conMatERR(1,1) \rightarrow errRate\{first \, 100 \, times \, iteration, gBestScore = inf\} ERR(it+1,1) \rightarrow errRate \,\&\, Constant \rightarrow ERR(it+1,1)\{not \, first \, 100 \, times \, iteration, gBestScore \neq inf\}$	Rewriting
f11	$Markergenes2(i, it+1) \rightarrow i, \{0 < errRate \leq constant, for \, 1 \leq i \leq 100\} f4, f5, f6, f7, f8, f9, f10$	Rewriting
f12	$Y(j,1) \rightarrow 0 \,\&\, max_j \rightarrow max_j - 1\{constant < errRate \,\&\, i \leq 50\} Y(j,1) \rightarrow 0 \,\&\, max_f \rightarrow max_f - 1\{constant < errRate \,\&\, i > 50\}$	Output
f13	$Y(j,1) \rightarrow +1\{1 \leq j \leq max_j * 3\} Y(j,1) \rightarrow -1\{max_j * 3 + 1 \leq j \leq max_j * 3 + max_f * 3\}$	Rewriting
f14	$size(wholedata) \rightarrow (M, N)\, A \rightarrow wholedata(1 : M-3, 1 : N)\, wholedata(M-3, N) \rightarrow 0$	Rewriting
f15	$wholedata \rightarrow A, k = k - 3$	Rewriting
f16	$a(i,j) \rightarrow \{(Particle3, it = 1), (Particle3, it = 2), ..., (Particle3, it = n)\}$	Division

From a numeric point of view, let assume that particle 1 is created to hold all the numeric value of dataset "a" for genes (Fig. 5). Executing the first part of the model (FS) will create another type of particle called particle 2, which holds sets of marker genes. The predefined "it" value indicates the number of particles type 2 to carry marker genes. Here we assumed "it" is 1. Executing the second part of the model FS/CL will create a different kind of particle called particle type 3. The number of these particles depends on the number of particle type 2. It means for each particle type 2, and there will be one counterpart as particle type 3. Particle type 3 keeps the updated genes resulted through executing FS/CL on the gene sets of particle type 2.

Indeed, particle type 2 and particle type 3 are the result of implementing division rule on particle type 1. It means that the whole dataset is divided into particle type two, which is resulted by executing the first part of the model

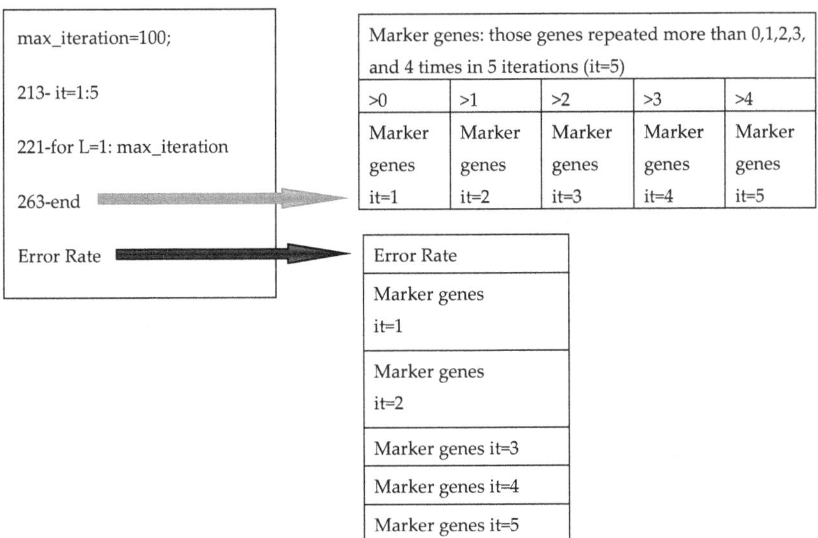

max_iteration=100;
213- it=1:5
221-for L=1: max_iteration
263-end
Error Rate
Marker genes: those genes repeated more than 0,1,2,3, and 4 times in 5 iterations (it=5)
>0
>1
>2
>3
>4
Marker genes it=1
Marker genes it=2
Marker genes it=3
Marker genes it=4
Marker genes it=5
Error Rate
Marker genes it=1
Marker genes it=2
Marker genes it=3
Marker genes it=4
Marker genes it=5

Example of the colorectal dataset:

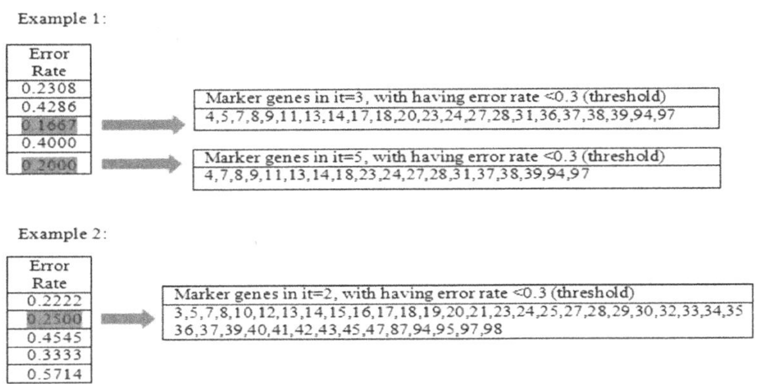

Example 1:
Error Rate
0.2308
0.4286
0.1667
0.4000
0.2000
Marker genes in it=3, with having error rate <0.3 (threshold)
4,5,7,8,9,11,13,14,17,18,20,23,24,27,28,31,36,37,38,39,94,97
Marker genes in it=5, with having error rate <0.3 (threshold)
4,7,8,9,11,13,14,18,23,24,27,28,31,37,38,39,94,97
Example 2:
Error Rate
0.2222
0.2500
0.4545
0.3333
0.5714
Marker genes in it=2, with having error rate <0.3 (threshold)
3,5,7,8,10,12,13,14,15,16,17,18,19,20,21,23,24,25,27,28,29,30,32,33,34,35
36,37,39,40,41,42,43,45,47,87,94,95,97,98

Fig. 3. Real data example from the colorectal dataset for kP-MObPSO.

(feature selection by MObPSO) and particle type three as the result of running the second part of computation (embedded feature selection and classification). These two types of particles (2 and 3) represent a different set of marker genes with varying rates of error, although particle type 3 is already updated based on particle type 2. Particle type 3 has already improved particle 2 in terms of error rate and ROC. This division rule itself can apply for any number of iterations.

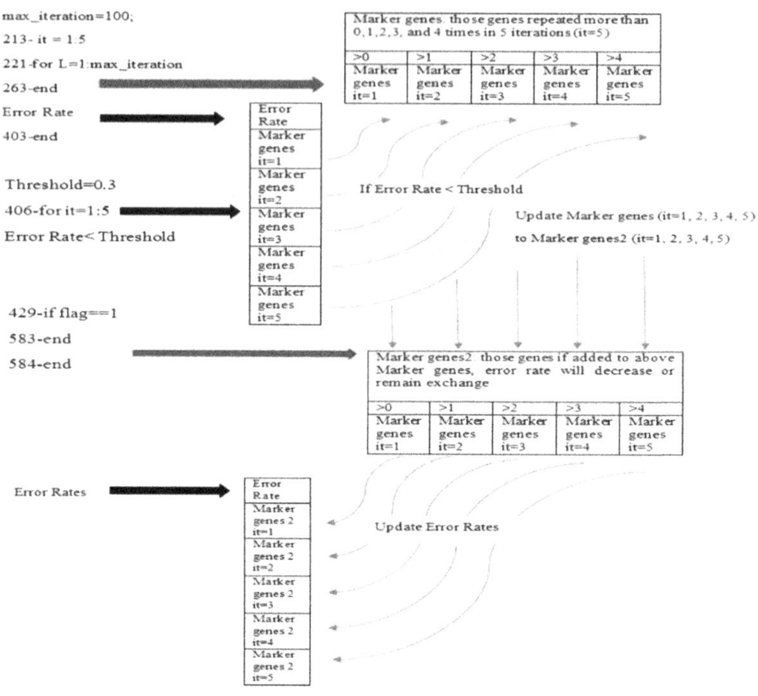

Example 1:

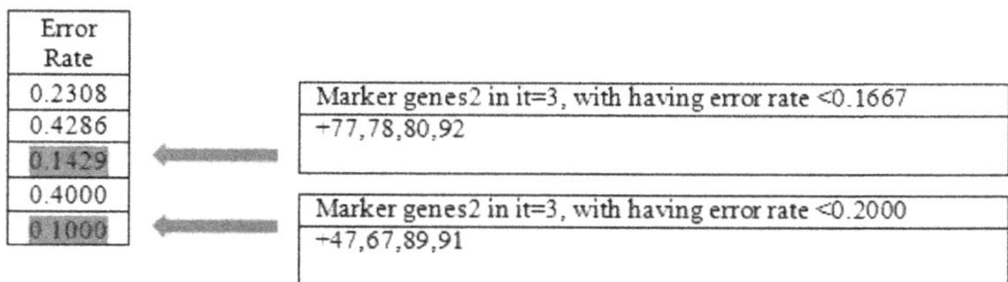

Error Rate
0.2308
0.4286
0.1429
0.4000
0.1000
Marker genes2 in it=3, with having error rate <0.1667
+77,78,80,92
Marker genes2 in it=3, with having error rate <0.2000
+47,67,89,91

Example 2:

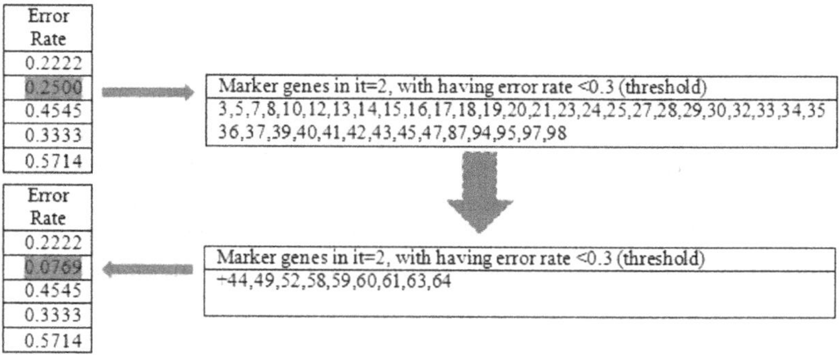

Fig. 4. Real data example from the colorectal dataset for embedded feature selection and classification kP-SVM.

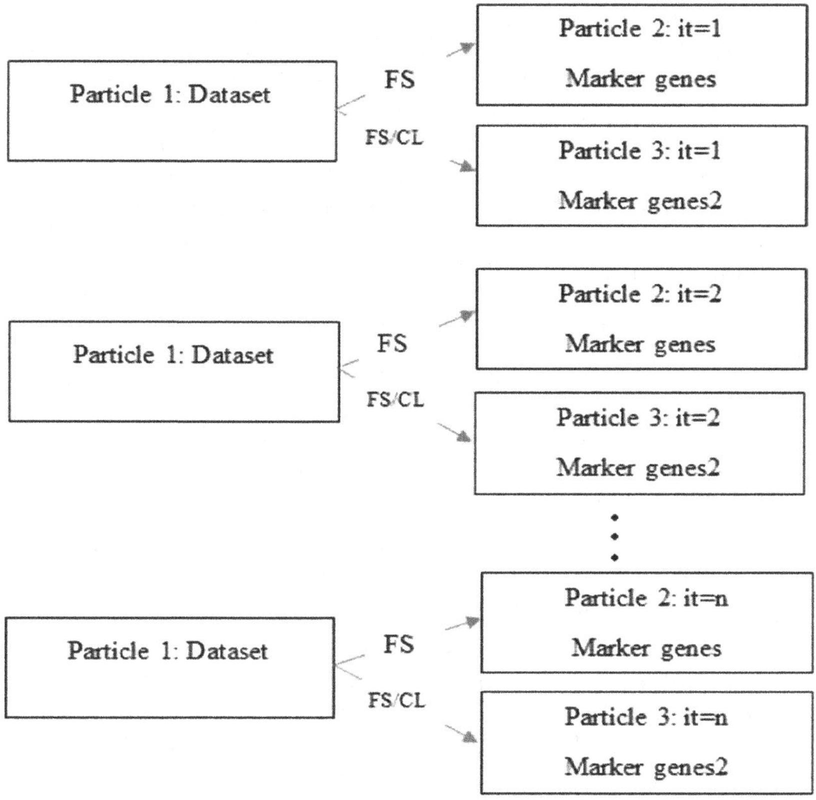

Fig. 5. Division rule applied on the initial dataset to divide genes into two type sets of genes.

4.1 Performance Evaluation Measurements

Measurements predicting the performance of a machine learning method based on inadequate data is difficult. Therefore, Cross-validation becomes the favorite when the researcher got a small amount of data. When machine-learning algorithms are used, decisions must be made on how to divide data for training and testing. To calculate the performance of machine learning methods, the entire data split training and testing sets, and 10-fold cross-validation, which is a famous method for evaluation, is applied afterward. The classification of data examined by a training set as forming a model, the verification of the model performed by the test set. The number of true negatives (TN), false negatives (FN), true positives (TP), and false positives (FP) are used to compute the efficiency of the classifier. The sensitivity and specificity are statistical measurements of checkout tests. Sensitivity states in the rate of the positive test result,

$$Sensitivity = TP/(TP + FN)100\% \tag{13}$$

Which has the following formula Accuracy shows the overall measure, which is: ROC represents the classifier performance without considering class distribution or error costs. For each algorithm, we also report the area under the ROC curve (AUC) value, which can indicate a model's balance ability between the TP rate and FP rate as a function of varying a classification threshold. As a result, we know if a model is biased to a particular class. An area of 1 represents a perfect classification, while an area of 0.5 represents a worthless model. There is another statistical measurement call F-measure; to evaluate characterization of the performance, it has the following formula.

$$Specificity = TN/(TN + FP) * 100\% \tag{14}$$

$$Accuracy = (TP + TN)/(TP + FP + TN + FN) * 100\% \tag{15}$$

$$F - measure = 2TP/(2TP + FP + FN) \tag{16}$$

4.2 Experimental Results

In this paper, a cell line dataset of colorectal cancer and breast cancer are downloaded from publicly available datasets in the Gene Expression Omnibus (GEO) repository (http://www.ncbi.nlm.nih.gov/gds).

Procedures and Data. Six samples of colorectal cancer and six samples of breast cancer are used according to Table 2 and 3, respectively that explains the specification of samples. When samples of a microarray dataset represent normal (benign) and cancer (malignant) tissue, classifying such samples is referred to as binary classification. Otherwise, when samples represent various subtypes of cancer, classification is referred to as multiclass cancer classification. In both cases, genes with significantly different expression in the two different classes (normal and tumor or two distinct subtypes of cancer) are labeled as differentially expressed genes. In this paper, we aimed to find these indicator genes, or marker genes, to distinguish normal and tumor genes (binary classification) in both the colorectal dataset and breast dataset.

According to Table 6, samples were provided from two subtypes, including subtype 1.1 and 2.1 with platform IDs: GPL570 (Affymetrix U133 Plus 2.0) and series GSE35896 to compare between 62 colorectal cancers. The first three samples, including GSM877130, GSM877141, and GSM877142, were from the subtype 1.1, and the latter three samples, GSM877127, GSM877138, and GSM877140, were from subtype 2.1. Before evaluating the model, we preprocessed the data. The number of genes in the raw data was 54676. A list of significant genes, including normal and tumor genes, were highlighted according to the results of a study [32] where the clinical, pathological, and biological features of different subtypes of colorectal tumors were compared. As a result, 382 tumor genes were selected according to the subtypes 1.1 and 2.1 in the initial cell-line dataset. The SNR was calculated according to the Eq. 13 for both normal and tumor genes. For each gene, the first three sets of samples were categorized as C1 (subtype 1.1), and the rest were classified as C2 (subtypes 2.1).

Table 6. Specification of colorectal cancer dataset.

Data Set Record GDS4379	Title: Colorectal cancer tumors	Summary: Analysis of primary colorectal cancer (CRC) tumors. CRC is a heterogeneous disease. Results provide insight into stratifying CRC tumor samples into subtypes and tailoring treatments for the CRC subtypes	Organism: Homo sapiens	Platform: GPL570: [$HG - U133\_Plus\_2$] Affymetrix Human Genome U133 Plus 2.0 Array
Citation: Schlicker A, Beran G, Chresta CM, McWalter G et al.	Subtypes of primary colorectal tumors correlate with response to targeted treatment in colorectal cell lines	BMC Med Genomics 2012 Dec 31; 5:66	PMID: 23272949	Platform ID: GPL570 Series: GSE35896, Gene expression data from 62 colorectal cancer
Samples	Cell line	Tissue	Disease state	genotype/variation
Subtype: 1.1				
GSM877130	CRC_42	Large Intestine	Adenocarcinoma	microsatellite.status: MSS
GSM877141	CRC_45	Large Intestine	Adenocarcinoma	microsatellite.status: MSS
GSM877142	CRC_61	Large Intestine	Adenocarcinoma	microsatellite.status: MSS
Subtype: 2.1				
GSM877127	CRC_01	Large Intestine	Adenocarcinoma	microsatellite.status: MSI
GSM877138	CRC_02	Large Intestine	Adenocarcinoma	microsatellite.status: MSI
GSM877140	CRC_19	Large Intestine	Adenocarcinoma	microsatellite.status: MSI

The SNR values were sorted in descending order, and 100 maximum SNR values (50 maximum SNR values for normal genes and 50 maximum SNR values for tumor genes) were selected for further analysis. Then the Signal-to-Noise Ratio (SNR) value (node weight) corresponding to each feature is calculated using mean and standard deviation (s.d.) of class l (cl) and class 2 (c2).

According to Table 7, six samples of breast cancer were used, with three of them related to papillary infiltrating ductal carcinoma and the rest belonging to a carcinosarcoma state of disease. Table 4 explains the specification of samples. Both samples were provided with platform IDs: GPL570 (Affymetrix U133 Plus 2.0) and series GSE3247 to compare cell lines from nine different cancer

Table 7. Specification of breast cancer dataset.

Age: 72	Sex: F	Epithelial: yes	Platform ID GPL570	Series (1) GSE32474
Comparison between cell lines from 9 different cancer tissue (NCI-60) (Affymetrix U133 Plus 2.0)	Treatment protocol	No treatment		
Samples	Cell line	Tissue	Disease state	Title
GSM803621	BT_549	breast	Papillary infiltrating ductal carcinoma	BR:BT_549 [113390hp133a11]
GSM803680	BT_549	breast	Papillary infiltrating ductal carcinoma	BR:BT_549 [113449hp133a11]
GSM803739	BT_549	breast	Papillary infiltrating ductal carcinoma	BR:BT_549 [118175hp133a11]
Age:74	Sex:F	Epithelial:yes	Source: Primary	Platform ID GPL570
Series (1) GSE32474	Comparison between cell lines from 9 different cancer tissue (NCI-60) (Affymetrix U133 Plus 2.0)	Treatment protocol	No treatment	
Samples	Cell line	Tissue	Disease state	Title
GSM803622	HS578T	breast	Carcinosarcoma	BR:HS578T [113391hp133a11]
GSM803681	HS578T	breast	Carcinosarcoma	BR:HS578T [113450hp133a11]
GSM803740	HS578T	breast	Carcinosarcoma	BR:HS578T [118176hp133a11]

tissues (NCI-60). First, three samples, including GSM803621, GSM803680, and GSM803739, were from the cell line BT_549, and the latter three samples, GSM803622, GSM803681, and GSM803740, were from cell line HS578T. Both the BT_549 and HS578T cell lines were derived from the claudin-low subtype of breast cancer, and classification aimed to distinguish between normal genes and tumor genes. Therefore, we sought to find these indicator genes, or marker genes, to determine normal and tumor genes (binary classification) in the claudin-low subtype of breast cancer.

Before evaluating the model, we preprocessed the data. The number of genes in the raw data was 4676. A list of significant genes, including normal and tumor genes, were highlighted according to the results of a study [32] where the clinical, pathological, and biological features of claudin-low tumors were compared to other subtypes of breast cancer, including luminal A, luminal B, HER2-enriched, and basal-like. As a result, 2329 tumor genes and 704 normal genes were selected in the initial cell-line dataset. The SNR was calculated according to the Eq. 4 for 3033 genes (including both normal and tumor genes). For each gene, the first three sets of samples were categorized as C1 (papillary infiltrating ductal carcinoma), and the rest were classified as C2 (carcinosarcoma). The SNR values were sorted in descending order, and 100 maximum SNR values (50 maximum SNR values for normal genes and 50 maximum SNR values for tumor genes) were selected for further analysis. During the final stage of preprocessing, the 100 genes were normalized according to Eq. 17. Therefore, all the values of the dataset (100 rows of genes times six sets of samples) were transformed to a value between 0 and 1 [0,1]. During the final stage of preprocessing, the 100 genes were normalized according to Eq. 17. For normalization, the minimum and maximum value of each gene (column) are calculated first. Then normalization is done. Where g_{ij} presents gene expression value of the i-th sample of the j-th gene and j presents gene expression values of j-th gene. Therefore, all the values of the dataset (100 rows of genes times six sets of samples) were transformed to a value between 0 and 1 [0,1].

$$Normalize(g_{ij}) = \frac{g_{ij} - minimum(g_{ij})}{maximum(g_{ij}) - minimum(g_{ij})} \tag{17}$$

The prepared data for the colorectal dataset and breast dataset was evaluated by the first part of our proposed model (kP-MObPSO) to search for gene markers. Then the resulted sets of gene markers were used as input for the second part of the proposed model (KP-Embedded feature selection and classification by SVM) to improve the set of genes through error rate. MATLAB R2014a was used for object-oriented coding of the entire proposed model. A KP rule-based SVM classifier proposed to evaluate the error rate of marker gene sets resulting from the first part of the proposed method (kP-MObPSO) and to evaluate the error rate of marker gene sets resulting from the second part of the proposed method (KP-Embedded feature selection and classification by SVM). The comparison between the error rates of marker genes sets from the first part and the second part of the proposed model indicates how the second part can find a better set of marker genes. To prove how good the second part of the model works to improve sets of marker genes, Weka 3.6.9 used to classify the marker genes resulted from the first part of the proposed model and second part of the model. Classification accuracy and ROC by Weka used to compare the first and second parts of the model. Using the Weka software, SMO (Implements John Platt's sequential minimal optimization algorithm for training a support vector classifier) works based on support vector machine theory. A linear SMO ($K(x, y) = <x, y>$), works to find the largest margin for separating hyperplane. It defines according to the aggregation of distances from hyperplane to the nearest positive or negative points. It is projected that as much as the margin is larger, the classifier will be more general. When the case is not detachable, the linear support vector machine looks for a trade-off that maximizes the margin and, on the other hand, minimizes errors.

5 Discussion

According to the results reported through the colorectal dataset, the comparison was made between the error rates of marker gene sets of proposed kP-MObPSO and the proposed embedded feature selection and classification method. C1 refers to normal gene sets, and C2 refers to cancerous gene sets. The threshold of error rate is given by 0.2000. The results indicate the second part of the proposed method to find a better group of marker genes due to better performance measures. For example, in one execution, the marker genes offered after feature selection through proposed kP-MObPSO are 4, 5, 7, 8, 9, 11, 13, 14, 17, 18, 20, 23, 24, 27, 28, 31, 36, 37, 38, 39, 94, 97 (Table 8) with the error rate 0.1667 (though classifier kP-SVM). The classification result on this marker gene by WEKA (Table 9) indicates this model is not reliable as the model just built on class C1 and ROC Area rate much less than 1 (ROC Area = 0.5). Adding marker genes as 77, 78, 80, 92 (Table 10) which have already found by the second part of the model shows better error rate 0.1429 (though classifier kP-SVM) and 0.0769 (through WEKA, Table 11) as well. According to Table 11, it also shows much more reliable modeling (accuracy = 92.30, ROC Area = 0.892, F-measure = 0.89 (Ave), recall = 0.89 (Ave), specificity = 0.833).

Table 8. Summary of the result for the colorectal dataset, kP-MObPSO feature selection/classifier kP-SVM.

Set no.	Marker genes	Error rate
1	4, 5, 7, 8, 9, 11, 13, 14, 17, 18, 20, 23, 24, 27, 28, 31, 36, 37, 38, 39, 94, 97	0.1667
2	4, 7, 8, 9, 11, 13, 14, 18, 23, 24, 27, 28, 31, 37, 38, 39, 94, 97	0.2000
3	3, 5, 7, 8, 10, 12, 13, 14, 15, 16, 17, 18, 19, 20, 21, 23, 24, 25, 27, 28, 29, 30, 32, 33, 34, 35, 36, 37, 39, 40, 41, 42, 43, 45, 47, 87, 94, 95, 97, 98	0.2500

In the second example, the marker genes offered after feature selection through proposed kP-MObPSO are 4, 7, 8, 9, 11, 13, 14, 18, 23, 24, 27, 28, 31, 37, 38, 39, 94, 97 (Table 8) with the error rate 0.2000 (Table 8). The classification result on this marker gene by WEKA (Table 9) again indicates this model is not reliable as the model just built on class C1 and ROC Area rate much less than 1 (ROC Area = 0.5). Adding marker genes as 47, 67, 89, 91 (Table 10) to the gene markers through the second part of the model improves the error rate to 0.1000 (though classifier KP-SVM) and 0.1364 (through WEKA, Table 11) as well. According to Table 11, it also shows much more reliable modeling (accuracy = 86.36%, ROC Area = 0.841, F-measure = 0.818(Ave), recall = 0.841(Ave), specificity = 0.800.

In the third example, the marker genes offered after feature selection through proposed kP-MObPSO are 3, 5, 7, 8, 10, 12, 13, 14, 15, 16, 17, 18, 19, 20, 21, 23, 24, 25, 26, 27, 28, 29, 30, 32, 33, 34, 35, 36, 37, 39, 40, 41, 42, 43, 45, 47, 87, 94, 95, 97, 98 (Table 8) with the error rate 0.2500. This set shows a reliable model as Table 9 through WEKA classifier (error rate = 0.1538, accuracy = 92.307, ROC Area = 0.892, F-measure = 0.89 (Ave), recall = 0.89 (Ave), specificity=20).

Table 9. Summary of the result for the colorectal dataset, kP-MObPSO feature selection/classifier kP-SVM.

Set No.	Error rate	Accuracy	ROC Area	F-measure	Sensitivity	Specificity
1	0.0909	90.90	C1 = 0.5 C2 = 0.5	C1 = 0.952 C2 = 0	C1 = 1 C2 = 0	C1 = 0 C2 = 0
2	0.1111	88.88	C1 = 0.5 C2 = 0.5	C1 = 0.941 C2 = 0	C1 = 1 C2 = 0	C1 = 0 C2 = 0
3	0.1538	92.307	C1 = 0.892 C2 = 0.892	C1 = 0.950 C2 = 0.830	C1 = 0.950 C2 = 0.830	C1 = 20 C2 = 20

Table 10. Summary of the result for colorectal dataset, Embedded kP-MObPSO feature selection/classifier kP-SVM.

Set No.	Marker genes	Error rate
1	4, 5, 7, 8, 9, 11, 13, 14, 17, 18, 20, 23, 24, 27, 28, 31, 36, 37, 38, 39, 94, 97, 77, 78, 80 ,92	0.1429
2	4 , 7, 8, 9, 11, 13, 14, 18, 23, 24, 27, 28, 31, 37, 38, 39, 94, 97, 47, 67, 89, 91	0.1000
3	3, 5, 7, 8, 10, 12, 13, 14, 15, 16, 17, 18, 19, 20, 21, 23, 24, 25, 27, 28, 29, 30, 32, 33, 34, 35, 36, 37, 39, 40, 41, 42, 43, 45, 47, 87, 94, 95, 97, 98, 44, 49, 52, 58, 59, 60, 61, 63, 64	0.0769

Table 11. Summary of the result for colorectal dataset, Embedded kP-MObPSO feature selection/classifier WEKA (SMO).

Set No.	Error rate	Accuracy	ROC Area	F-measure	Sensitivity	Specificity
1	0.0769	92.30%	C1 = 0.892 C2 = 0.892	C1 = 0.950 C2 = 0.833	C1 = 0.950 C2 = 0.833	C1 = 0.833 C2 = 0.833
2	0.1364	86.36%	C1 = 0.841 C2 = 0.841	C1 = 0.909 C2 = 0.727	C1 = 0.882 C2 = 0.800	C1 = 0.800 C2 = 0.800
3	0.0408	95.91%	C1 = 0.973 C2 = 0.973	C1 = 0.972 C2 = 0.923	C1 = 0.946 C2 = 1	C1 = 1 C2 = 1

Adding marker genes as 44, 49, 52, 59, 61, 63, 64 (Table 10) to the gene markers through the second part of the model improves the error rate to 0.0769 (though classifier KP-SVM) and 0.0408 (through WEKA, Table 11) as well. According to Table 11, it also shows much more reliable modeling (accuracy = 95.91, ROC Area = 0.973, F-measure = 0.947(Ave), recall = 0.973(Ave), specificity = 1).

According to the result of the second part of the model, all three sets of marker genes show good stability, so in terms of model reliability, all have the same priority. As far as the error rate is considered, the lowest error rate (0.0408) belongs to the set that consists of a higher number of genes as marker genes, while the second good error rate (0.0769) consists of just 24 genes as marker genes. Thus, it can be a trade-off between the lowest error rate and the lowest number of marker genes. When the lowest error rate is considered, the third model can be chosen, and when the goal is to achieve the lowest number of genes with a good error rate, the second model is the best choice.

According to the results reported through the h breast cancer dataset, the comparison done between the performance measures of marker genes sets from the first part and the second part of the proposed model. Again, C1 refers to normal gene sets, and C2 refers to cancerous gene sets. The threshold of error rate is given by 0.4000. The results indicate the second part of the proposed model to find a better set of marker genes due to better and reliable performance measures.

Table 12. Summary of the result for Breast dataset, kP-MObPSO feature selection/classifier kP-SVM.

Set No.	Marker genes	Error rate
1	6, 8, 9, 10, 11, 12, 13, 14, 15, 16, 17, 19, 21, 22, 23, 26, 27, 28, 30, 31, 32, 33, 34, 36, 37, 38, 39, 40, 41, 46, 49, 63, 69, 75, 77, 81, 82, 87, 88, 93, 94, 95, 97, 99, 100	0.3846
2	6, 8, 13, 14, 17, 19, 20, 22, 23, 29, 31, 32, 33, 34, 35, 36, 37, 38, 39, 40, 42, 44, 45, 79, 81, 83, 84, 86, 92, 93, 96, 97, 98, 99, 100	0.2000
3	7, 8, 13, 14, 22, 23, 24, 27, 29, 31, 32, 35, 38, 44, 45, 46, 49, 74, 75, 76, 79, 80, 82, 87, 88, 91, 92, 94, 97, 99	0.1111

For example, in one execution, the marker genes offered after feature selection through proposed kP-MObPSO are 6, 8, 9, 10, 11, 12, 13, 14, 15, 16, 17, 19, 21, 22, 23, 26, 27, 28, 30, 31, 32, 33, 34, 36, 37, 38, 39, 40, 41, 46, 49, 63, 69, 75, 77, 81, 82, 87, 88, 93, 94, 95, 97, 99, 100 (Table 12) with the error rate 0.3846 (though classifier KP-SVM) and 0.2000 (through WEKA, Table 13). The classification result on this marker gene by WEKA (Table 13) indicates that the model is reliable as (accuracy=80%, ROC Area = 0.816, F-measure = 0.784 (Ave), recall = 0.815 (Ave), specificity = 0.857). Adding marker genes as 5, 7, 18, 59, 66, and 96 (Table 14) which have already found by second part of the model shows better error rate 0.3333 (though classifier KP-SVM) and 0.1923 (through WEKA, Table 15) respectively. Adding these all 6 genes also shows much stronger model as Table 15 (accuracy = 80.76%, ROC Area = 0.827, F-measure = 0.797 (Ave), recall = 0.826 (Ave), specificity = 0.882).

In another execution, the marker genes offered after feature selection through proposed kP-MObPSO are 6, 8, 13, 14, 17, 19, 20, 22, 23, 29, 31, 32, 33, 34, 35, 36, 37, 38, 39, 40, 42, 44, 45, 79, 81, 83, 84, 86, 92, 93, 96, 97, 98, 99, 100 (Table 12) with the error rate 0.2000 (though classifier KP-SVM) and 0.3871

Table 13. Summary of the result for Breast dataset, kP-MObPSO feature selection/classifier WEKA (SMO).

Set No.	Error rate	Accuracy	ROC Area	F-measure	Sensitivity	Specificity
1	0.2000	80	C1 = 0.816 C2 = 0.816	C1 = 0.842 C2 = 0.727	C1 = 0.774 C2 = 0.857	C1 = 0.857 C2 = 0.857
2	0.3871	61.29	C1 = 0.413 C2 = 0.413	C1 = 0.76 C2 = 0	C1 = 0.826 C2 = 0	C1 = 0 C2 = 0
3	0.1667	83.33	C1 = 0.835 C2 = 0.835	C1 = 0.848 C2 = 0.815	C1 = 0.824 C2 = 0.846	C1 = 3.66 C2 = 3.66

(through WEKA, Table 13). According to Table 13, it doesn't show a stable model because the model just created on C1 genes due to ROC = 0.413. With adding genes number 53 and 77 (Table 14) through the embedded part of the model, error rate decreased to 0.1111 (though classifier KP-SVM) and 0.2162 (through WEKA, Table 15). Adding these 2 genes shows reliable and stronger model as Table 15 (accuracy = 78.37%, ROC Area = 0.784, F-measure = 0.818 (Ave), recall = 0.784 (Ave), specificity = 0.785).

Table 14. Summary of the result for Breast dataset, Embedded kP-MObPSO feature selection/classifier kP-SVM.

Set No.	Marker genes	Error rate
1	6, 8, 9, 10, 11, 12, 13, 14, 15, 16, 17, 19, 21, 22, 23, 26, 27, 28, 30, 31, 32, 33, 34, 36, 37, 38, 39, 40, 41, 46, 49, 63, 69, 75, 77, 81, 82, 87, 88, 93, 94, 95, 97, 99, 100, 5, 7, 18, 59, 66, 96	0.3333
2	6, 8, 13, 14, 17, 19, 20, 22, 23, 29, 31, 32, 33, 34, 35, 36, 37, 38, 39, 40, 42, 44, 45, 79, 81, 83, 84, 86, 92, 93, 96, 97, 98, 99, 100, 53, 77	0.1111
3	same set of marker genes	0.1111

In the third execution, the marker genes offered after feature selection through proposed kP-MObPSO are 7, 8, 13, 14, 22, 23, 24, 27, 29, 31, 32, 35, 38, 44, 45, 46, 49, 74, 75, 76, 79, 80, 82, 87, 88, 91, 92, 94, 97, 99 (Table 12) with the error rate 0.1111 (though classifier KP-SVM) and 0.1667 (through WEKA, Table 13). According to the Table 13, this set makes a reliable model as (accuracy = 83.33%, ROC Area = 0.835, F-measure = 0.831 (Ave), recall = 0.835 (Ave), specificity = 3.66). The embedded part of our proposed model couldn't find any gene that can improve this set of marker genes in terms of error rate and stability.

The results of the second part of the model indicate the first marker gene set with 51 genes (Table 14) with accuracy = 80.76 and ROC = 0.827 (Table 15) can be a candidate set. Also, the third set of marker genes with 30 genes (Table 12)

Table 15. Summary of the result for Breast dataset, Embedded kP-MObPSO feature selection/classifier WEKA (SMO).

Set No.	Error rate	Accuracy	ROC Area	F-measure	Sensitivity	Specificity
1	0.1923	80.76	C1 = 0.827 C2 = 0.827	C1 = 0.844 C2 = 0.750	C1 = 0.771 C2 = 0.882	C1 = 0.882 C2 = 0.882
2	0.2162	78.37	C1 = 0.784 C2 = 0.784	C1 = 0.818 C2 = 0.818	C1 = 0.783 C2 = 0.786	C1 = 0.785 C2 = 0.785
3	No improvement found	No improvement found	No improvement found	No improvement found	No improvement found	No improvement found

with accuracy = 83.33 and ROC = 0.835 (Table 13) shows a quite reliable model. It is clear that in comparison with the first marker gene set (Table 14), the third set of marker genes (Table 12) is the best choice in terms of performance measures unless there is a trade-off to have more marker genes as an indicator.

According to Table 16, implementing pure MObPSO on the colorectal dataset and breast dataset leads to an accuracy of 87.50% and 78.57%, respectively. It is while the proposed embedded feature selection and classification method got the accuracy of 92.30%, 86.36%, and 95.91% (Table 11) in a colorectal dataset in comparison to 87.50% in pure MObPSO. In the breast dataset, the proposed embedded feature selection and classification method got an accuracy of 80.76% and 78.37% (Table 15) in comparison to 78.57% in pure MObPSO.

Table 16. Specification of colorectal cancer dataset.

Colorectal dataset				
Accuracy	ROC Area	F-measure	Sensitivity	Specificity
87.50%	0.90	0.877	0.875	0.906
Breast dataset				
Accuracy	ROC Area	F-measure	Sensitivity	Specificity
78.57%	0.72	0.785	0.786	0.792

6 Conclusions

Feature selection is a commonly addressed problem in classification. In gene expression-based cancer classification, a large number of genes, in conjunction with a small number of samples, make the gene selection problem more critical but also more challenging.

In this paper, we presented a method which, at a higher level, is the mix of membrane computing with an intelligent algorithm to increase classification performance with microarray data. The proposed model consists of two main parts. First, kP-MObPSO, which resembles a wrapper type feature selection, and the second part that improves the results of the first part through an embedded feature selection and classification idea based on kP system. The experiment aimed to evaluate the set of marker genes in terms of performance measures that can interpret based on accuracy, ROC, error rate, recall, F-measure, specificity, and the size of the marker gene set.

The proposed model generates genes with better performance measures. The model based on the kP system resembles the interaction of genes between and inside compartments. The most important feature of the kP computation system has the computation rules as division, rewriting/ communication, and input/output that has the nature of parallelism and non-determinism. In this paper, the rules of the proposed model are implemented in a sequential type of execution. We have also implemented the model in parallel through multi-cores and GPU programming [33–35].

Acknowledgements. The efforts of grant for Development of Membrane Computing Software (Universiti Kebangsaan Malaysia (UKM), UKM Grant Code: GGP-2019-023) has been acknowledged, as this support has played a vital role in the successful accomplishment of the research.

References

1. Păun, G.: Computing with membranes. J. Comput. Syst. Sci. **61**(1), 108–143 (2000)
2. Zhang, G., Haina, R., Ferrante, R., Pérez-Jiménez, M.J.: An optimization spiking neural P system for approximately solving combinatorial optimization problems. Int. J. Neural Syst. **24**(5), 1440006 (2014)
3. Huang, L., Wang, N.: An optimization algorithm inspired by membrane computing. In: Jiao, L., Wang, L., Gao, X., Liu, J., Wu, F. (eds.) ICNC 2006. LNCS, vol. 4222, pp. 49–52. Springer, Heidelberg (2006). https://doi.org/10.1007/11881223_7
4. Frisco, P., Corne, D.W.: Modeling the dynamics of HIV infection with Conformon-P systems and cellular automata. In: Eleftherakis, G., Kefalas, P., Păun, G., Rozenberg, G., Salomaa, A. (eds.) WMC 2007. LNCS, vol. 4860, pp. 21–31. Springer, Heidelberg (2007). https://doi.org/10.1007/978-3-540-77312-2_2
5. Gutiérrez-Naranjo, M.A., Pérez-Jiménez, M.J., Romero-Campero, F.J.: Simulating avascular tumors with membrane systems. In: Proceedings of the Third Brainstorming Week on Membrane Computing, pp. 185–196. Fénix Editora, Sevilla (Spain) (2005)
6. Pérez-Jiménez, M.J., Romero-Campero, F.J.: A study of .the robustness of the EGFR signalling cascade using continuous membrane systems. In: Mira, J., Álvarez, J.R. (eds.) IWINAC 2005. LNCS, vol. 3561, pp. 268–278. Springer, Heidelberg (2005). https://doi.org/10.1007/11499220_28
7. Bernardini, F., Gheorghe, M., Krasnogor, N.: Quorum sensing P systems. Theor. Comput. Sci. **371**(1), 20–33 (2007)
8. Muniyandi, R.C., Zin, A.M., Sanders, J.: Converting differential-equation models of biological systems to membrane computing. BioSystems **114**(3), 219–226 (2013)
9. Siegel, R., DeSantis, C., Jemal, A.: Colorectal cancer statistics. CA: Cancer J. Clin. **64**(2), 104–117 (2014)
10. Gheorghe, M., Ipate, F., Dragomir, C., Mierla, L., Valencia-Cabrera, L., Garcia-Quismondo, M., Pérez-Jiménez, M.J.: Kernel P Systems Version I. In: Proceedings of the Eleventh Brainstorming Week on Membrane Computing, pp. 97–124. Fénix Editora, Sevilla(Spain) (2013)

11. Mohapatra, P., Chakravarty, S.: Modified PSO based feature selection for microarray data classification. In: Proceedings of the 2015 IEEE Power, Communication and Information Technology Conference (PCITC). IEEE, Bhubaneswar (India) (2015)
12. Kar, S., Sharma, K.D., Maitra, M.: Gene selection from microarray gene expression data for classification of cancer subgroups employing PSO and adaptive K-nearest neighborhood technique. Expert Syst. Appl. **42**(1), 612–627 (2015)
13. Chinnaswamy, A., Srinivasan, R.: Hybrid feature selection using correlation coefficient and particle swarm optimization on microarray gene expression data. In: Snášel, V., Abraham, A., Krömer, P., Pant, M., Muda, A.K. (eds.) Innovations in Bio-Inspired Computing and Applications. AISC, vol. 424, pp. 229–239. Springer, Cham (2016). https://doi.org/10.1007/978-3-319-28031-8_20
14. Mandal, M., Mukhopadhyay, A.: A graph-theoretic approach for identifying non-redundant and relevant gene markers from microarray data using multi-objective binary PSO. PloS One **9**(3), e90949 (2014)
15. Apolloni, J., Leguizamón, G., Alba, E.: Two-hybrid wrapper-filter feature selection algorithms applied to high-dimensional microarray experiments. Appl. Soft Comput. **38**, 922–932 (2016)
16. Elyasigomari, V., Mirjafari, M.S., Screen, H.R.C., Shaheed, M.H.: Cancer classification using a novel gene selection approach by means of shuffling based on data clustering with optimization. Appl. Soft Comput. **35**, 43–51 (2015)
17. Sheikhpour, R., Sarram, M.A., Sheikhpour, R.: Particle swarm optimization for bandwidth determination and feature selection of kernel density estimation-based classifiers in diagnosis of breast cancer. Appl. Soft Comput. **40**, 113–131 (2016)
18. Duan, K., Rajapakse, J.C.: A variant of SVM-RFE for gene selection in cancer classification with expression data. In: Proceedings of the 2004 IEEE Symposium on Computational Intelligence in Bioinformatics and Computational Biology (CIBCB). IEEE, La Jolla (USA) (2004)
19. Tang, Y., Zhang, Y.Q., Huang, Z.: Development of two-stage SVM-RFE gene selection strategy for microarray expression data analysis. IEEE/ACM Trans. Comput. Biol. Bioinform. **4**(3), 365–381 (2007)
20. Huerta, E.B., Montiel, A.H., Caporal, R.M., Lopez, M.A: Hybrid framework using multiple-filters and an embedded approach for an efficient and robust selection and classification of microarray data. IEEE/ACM Trans. Comput. Biol. Bioinform **13**(1), 12–26 (2015)
21. Pashaei, E., Ozen, M., Aydin, N.: Gene selection and classification approach for microarray data based on Random Forest Ranking and BBHA. In: Proceedings of the 2016 IEEE-EMBS International Conference on Biomedical and Health Informatics (BHI). IEEE, Las Vegas (USA) (2016)
22. Shapiro, G.P., Tamayo, P.: Microarray data mining: facing the challenges. ACM SIGKDD Explor. Newslett. **5**(2), 1–5 (2003)
23. Hall, M.A.: Correlation-based feature selection for machine learning. The University of Waikato, Hamilton (New Zealand) (1999). https://www.cs.waikato.ac.nz/~mhall/thesis.pdf. Accessed 20 July 2020
24. Koller, D., Sahami, M.: Toward optimal feature selection. In: Proceedings of the Thirteenth International Conference on International Conference on Machine Learning (ICML), pp. 284–292. Morgan Kaufmann Publishers Inc., San Francisco (USA) (1996)
25. Yu, L., Liu, H.: Efficient feature selection via analysis of relevance and redundancy. J. Mach. Learn. Res. **5**, 1205–1224 (2004)

26. Furey, T.S., Cristianini, N., Duffy, N., Bednarski, D.W., Schummer, M., Haussler, D.: Support vector machine classification and validation of cancer tissue samples using microarray expression data. Bioinformatics **16**(10), 906–914 (2000)
27. Lin, S.W., Ying, K.C., Chen, S.C., Lee, Z.J.: Particle swarm optimization for parameter determination and feature selection of support vector machines. Expert Syst. Appl. **35**(4), 1817–1824 (2008)
28. Rahman, M.A., Muniyandi, R.C.: An enhancement in cancer classification accuracy using a two-step feature selection method based on artificial neural networks with 15 neurons. Symmetry **12**, 271 (2020)
29. Scholkopf, B., Guyon, I., Weston, J.: Statistical Learning and Kernel Methods in Bioinformatics. IOS Press, Amsterdam (2003)
30. Guyon, I., Weston, J., Barnhill, S., Vapnik, V.: Gene selection for cancer classification using support vector machines. Mach. Learn. **46**(1–3), 389–422 (2002). https://doi.org/10.1023/A:1012487302797
31. Burges, C.J.: A tutorial on support vector machines for pattern recognition. Data Min. Knowl. Discov. **2**(2), 121–167 (1998)
32. Schlicker, A., et al.: Subtypes of primary colorectal tumors correlate with response to targeted treatment in colorectal cell lines. BMC Med. Genomics **5**(1), 66 (2012). https://doi.org/10.1186/1755-8794-5-66
33. Elkhani, N., Muniyandi, R.C., Zhang, G.: Multi-objective binary PSO with kernel P system on GPU. Int. J. Comput. Commun. Control **13**(3), 323–336 (2018)
34. Elkhani, N., Muniyandi, R.C.: A multiple core execution for multiobjective binary particle swarm optimization feature selection method with the kernel P system framework. J. Optimiz. **13**, 1–14 (2017)
35. Muniyandi, R.C., Maroosi, A.: A representation of membrane computing with a clustering algorithm on the graphical processing unit. Processes **8**(9), 1199 (2020)

Evolutionary P Systems: The Notion and an Example

Taishin Y. Nishida(✉)

Department of Information Systems, Toyama Prefectural University, 5180 Kurokawa, Imizu, Toyama 939-0398, Japan
nishida@pu-toyama.ac.jp

Abstract. We set up the notion of evolutionary P systems as P systems with description of rules, or genomes, and translation, evaluation, selection, and modification operators on genomes. The system has a possibility of evolving a desired function. We propose a tissue evolutionary P system which evolves a context-free grammar generating a given target language, i.e., an evolutionary P system for grammatical inference. Experiments show that the proposed systems can evolve some context-free grammars generating the language $\{a^n b^n \mid n > 0\}$ and the Dyck language over $\{a, b\}$.

1 Introduction

Since P systems are computational models incorporating various phenomena from natural cells, there should be P systems which evolve and obtain some functions. Some authors have noticed to the relation between P systems and evolution or adaption [1,15]. This work is a continuation of [8]. First we set up a general framework of evolutionary P systems, which consists of a P system with genomes and evaluation, selection, and modification operators on genomes. Like living organisms, a genome is a sequence of basic objects, which should be symbols in the formal models. In evolution, a genome is translated to a collection of rule sets of a P system, the P system computes some results using the rules, and finally the results are scored to setup a fitness value for the genome. Selection and modification operators can be borrowed from genetic algorithm.

Now one might think that evolutionary P systems are contained to the family of membrane algorithms [3,6,9,16–18] because some type of membrane algorithm is a mixture of genetic algorithm and P system. But it should be noted that evolutionary P systems and membrane algorithms are different branches of P systems. Membrane algorithms are frameworks for membrane systems to be solvers of real problems, especially NP hard optimizing problems. Membrane algorithms can use wide variety of algorithms for optimizing problems, in addition to genetic algorithms [7]. On the other hand, evolutionary P systems are, as described above, P systems with evolution, in which rules are described in strings (genomes) and change. Evolutionary P systems can model biological evolution, evolution in artificial life, or function acquisition or automatic learning for industrial applications.

R. Freund et al. (Eds.): CMC 2020, LNCS 12687, pp. 126–134, 2021.
https://doi.org/10.1007/978-3-030-77102-7_7

Second, as an example of evolutionary P systems, we give an improved, compared to the system given in [8], evolutionary P system for grammatical inference. If the genomes are translated to string rewriting rules of context-free type and evaluated under the criterion how similar language to a given target language is generated, then the system evolve to a grammar which generates the target language, i.e., it is a grammatical inference system. In this paper, we develop a tissue evolutionary P system for grammatical inference.

Grammatical inference (GI) has a considerably long history. Some basic theoretical results go back to 1960's [2]. But GI has contemporary meaning because there are many applications of GI, including software engineering, domain-specific languages, etc., [5,13]. Since there are no known effective and deterministic algorithms for inferring a context-free grammar, researchers are interested in GI algorithms using meta-heuristics or soft computing [4,10,14]. Evolutionary P system for GI might be a new algorithm of this class.

This paper organized as follows. In Sect. 2 formal framework of evolutionary P systems is given. Tissue evolutionary P system for GI is described in Sect. 3. The details of the experiments of the system are found in Sect. 4. The last section contains discussions.

2 Definition of Evolutionary P Systems

As mentioned in Introduction, an evolutionary P system has a description of rules, or a genome, a mechanism of translating the genome to its functionality, an evaluating operator which scores the genome under some criteria to attach a fitness value, a selection operator which selects the genome according to the fitness values, and a modifying operator which modifies genomes. In this section we give a formal framework of evolutionary P systems. The readers are assumed to be familiar with basics of formal language theory and P systems [11,12].

First we give a unified notation of a cell-like, tissue, or neural "rule-fixed" P system of degree k by the construction

$$\Pi = (O, \mu, w_1, \ldots, w_k, in, out, \mathcal{R})$$

where:

- O is an alphabet of objects.
- $\mu = (V, E)$ is the membrane structure with $V = \{1, \ldots, k\}$ and $E \subseteq V \times V - \{(i,i) \mid i \in V\}$.
- $w_1, \ldots, w_k$ are the initial multisets of regions 1, ..., k, respectively.
- $in, out \in \{1, \ldots, k\}$ are the input and output regions, respectively.
- $\mathcal{R}$ is a collection of sets of rules.

Each variant of P system gives specific conditions on the above elements. For example, μ is a tree and $\mathcal{R} = (R_1, \ldots, R_k)$ where $R_i \subseteq O^+ \times (O \times targets)^*$ for a cell-like P system with multiset rewriting rules and μ is a graph and $\mathcal{R}$ is a set of symport/antiport rules associated to each edge in μ for a tissue P system with symport/antiport rules. It should be noted that this simple notation cannot describe all variants of P systems. The notation is intended to describe typical variants of rule-fixed P systems. We introduce a "rule-less" P system, or *membrane skeleton*, by dropping $\mathcal{R}$, that is,

$$\Pi_0 = (O, \mu, w_1, \ldots, w_k, in, out)$$

is a membrane skeleton.

Now we move to evolutionary P systems. First we give the notation of a *genome* P system of degree k which has genomes instead of rules

$$\Pi = (\Pi_0, \Gamma, u_1^{(0)}, \ldots, u_p^{(0)}, \phi)$$

where:

- $\Pi_0 = (O, \mu, w_1, \ldots, w_k, in, out)$ is a membrane skeleton.
- Γ is the alphabet of genomes satisfying $\Gamma \cap O = \emptyset$.
- $u_1^{(0)}, \ldots, u_p^{(0)} \in \Gamma^*$ are the initial population of genomes, where $p > 0$ is the population size.
- ϕ is a translating function which maps a genome and the membrane structure to a collection of sets of rules, that is, $\phi(u_i, \mu) = \mathcal{R}_i$ is a collection of rules for $i \in \{1, \ldots, p\}$.

Example: Let $\Gamma = \bar{O} \cup \{h, o\} \cup \{i_j \mid j \in \{1, \ldots, k\}\} \cup \{\#\} \cup \{r_i \mid i \in \{1, \ldots, k\}\}$ where $\bar{O}$ is a disjoint copy of O. Let $D = \{h, o\} \cup \{i_j \mid j \in \{1, \ldots, k\}\}$ and let f be the mapping $f : \bar{O}D \to O \times targets$ defined by

$$f(\bar{a}t) = \begin{cases} (a, here) & \text{if } t = h \\ (a, out) & \text{if } t = o \\ (a, in_j) & \text{if } t = i_j \end{cases}.$$

where $\bar{a} \in \bar{O}$ and $t \in D$. We note that f is bijective and that f can be extended to a homomorphism from $(\bar{O}D)^*$ to $(O \times targets)^*$. A multiset rewriting rule $x \to y$ in region i is coded by $r_i \bar{x} \# f^{-1}(y)$. The above coding defines the translation function ϕ by

$$\phi(u, \mu) = \mathcal{R} = (R_1, \ldots, R_k)$$

where R_i ($i \in \{1, \ldots, k\}$) is the set

$$R_i = \{x \to y \mid u = \alpha r_i \bar{x} \# f^{-1}(y)\beta, \alpha, \beta \in \Gamma^*, \beta = \lambda \text{ or begin with } r_j\}.$$

We note that in this example ϕ does not depend on the membrane structure μ because the set of multiset rewriting rules is attached to a region. □

An evaluation operator will entirely depend on an interpretation of objects and rules. For example, if the rules compute a solution of an optimization problem, then a goodness measure with respect to the optimizing criteria will be the fitness value of the rules or the genome. If the set of rules reproduced the cell from elementary material, then the efficiency of the reproduction would be the fitness and then the system would be a model of artificial life. Anyway, evaluation starts from a population of P systems

$$\Pi_i = (\Pi_0, \phi(u_i, \mu))$$

for $i \in \{1, \ldots, p\}$, where u_i is the i-th genome in current generation. Every Π_i starts computation and gets some fitness value by the evaluation criteria.

Once evaluation operator is established, selection and modification operators on genomes will be resemble to those operators from genetic algorithm. Selection and modification make the population of genomes of the next generation. Evaluation, selection, and modification operators play the central roll in evolution and we denote the triplet by ϵ.

Now the pair (Π, ϵ) forms an *evolutionary P system*, or *EP system*. An EP system alternates an evaluation phase and a selection-modification phase when it develops. In evaluation phase the system translates current genomes to a population of P systems. Each P system proceeds normal computation and get a result. Finally the result is evaluated to get a fitness value. In selection-modification phase, genomes are selected by the fitness values and genomes are modified to the next generation. As mentioned above, this phase resembles to genetic algorithm.

3 Evolutionary P System for Grammatical Inference

In what follows we concentrate an EP system whose genomes are translated to string rewriting rules. The rules have the form of context-free production rules, that is, terminal symbols are discriminated from nonterminal symbols and every left-hand side of a rule consists of one nonterminal. The initial string is the start symbol. Thus each cell has a context-free grammar. If a set of sample words from a target language L_t is given, the grammar in a cell is evaluated by the next scheme:

- Every word in the sample set is parsed using the grammar.
- If the parsing is success, then a highest value is added to the fitness.
- If the parsing fails, then the "badness" of the parsing is measured (e.g., a distance between the target word and the partial word generated) and a corresponding value is added to the fitness.

The above evaluation scores the highest fitness to a trivial grammar generating Σ^* where Σ is the alphabet of L_t. To avoid this, the grammars must be evaluated under a set of negative sample words from $\Sigma^* - L_t$. An appropriate combination of positive and negative evaluations, selections, and modifications may evolve a grammar which generates the target language L_t. Now the total EP system becomes a grammatical inference system.

In [8], a single cell-like P system with degree 3 was used as the membrane skeleton of the EP system for grammatical inference. In this study, we introduce a tissue P system which consists of a population of membrane skeletons developed in [8]. We call a membrane skeleton a cell. The highest fitness value among genomes in a cell is assumed to be the fitness of the cell.

There is a communication channel between every pair of different cells. The cells in the tissue P system communicate in the next manner:

- Every cell but the best (w.r.t. fitness) cell get a randomly selected genome from the best cell and the genome replaces with a randomly selected genome in the cell.
- If a cell dies, i.e., if every genome in the cell has fitness 0, then a new cell which consists of randomly selected genomes from other cells is created and is replaced with the dead cell.

The communication rules are fixed throughout the computation, that is, the communications are a part of selection-modification phase and should not be coded in the genome. These communications are expected to keep variety of genomes and to avoid falling in a local minima. It should be noted that the communications are a natural consequence of P system because P system provides a wide variety of phenomena of the biological cells, in this case, tissue structure and communications among cells.

We use iteration steps as the terminate condition, that is, evolution terminates after a given iteration steps. The initial genomes are set to random sequences. The evolution in the P system proceeds as follows:

Initialization.
While step < max-iterations.
  Evaluation.
  Selection.
  Crossover and mutation.
  Migration.
  step++.

4 Experiments

We select three target languages:

- $L_1 = \{a^n b^n \mid n \geq 1\}$. We denote L_1 by $a^n b^n$.
- $L_2 = \{ww^R \mid w \in \{a,b\}^+\}$, where w^R is the inverse of w. We denote L_2 by ww^R.
- $L_3 =$ (the Dyck language over $\{a,b\}$). We denote L_3 by Dyck.

We use two nonterminal symbols A and S in which S is the start symbol. For the genome sequence, we introduce the alphabet $\{\tilde{A}, \tilde{S}, \bar{A}, \bar{S}, \bar{a}, \bar{b}\}$. A rule $X \to \alpha$ is represented by the string $\tilde{X}\bar{\alpha}$. A genome string consists of a concatenation of the rule string of the form $\tilde{X}\bar{\alpha}$.

In the evaluation phase, every translated grammar tries to parse the (positive or negative) target words by the $LL(1)$ parsing. If the translated grammar is not $LL(1)$, then more than one rules are applicable for a pair of a terminal and a nonterminal. In this case one rule is selected randomly. For a target word w_t, the parser generates some word w_g even if it fails. The Hamming distance $d(w_t, w_g)$ is calculated by

$$d(w_t, w_g) = \sum_{i=1}^{s} \delta((w_t)_i, (w_g)_i) + \frac{1}{2}||w_t| - |w_g||$$

where w_i represents the i-th letter of a word w, $s = \min(|w_t|, |w_g|)$, and $\delta(x, y)$ is given by

$$\delta(x, y) = \begin{cases} 0, & x = y \\ 1, & x \neq y \end{cases} .$$

A target word w_t from the positive samples gives the fitness value

$$\frac{3}{2}|w_t| - d(w_t, w_g)$$

or 0 if $\frac{3}{2}|w_t| < d(w_t, w_g)$. If $w_g = w_t$, then the maximum value $\frac{3}{2}|w_t|$ is added to the fitness. For a target from the negative samples, two types of evaluation are examined. The first type (negative 1) adds the value $d(w_t, w_g)$ to the fitness value and genomes are selected in the same manner as the positive samples. The other type (negative 2) does not compute fitness value for the negative samples but a genome whose translated grammar generates a negative target word (generates w_g with $d(w_t, w_g) = 0$) is erased. This method was inspired by the immune-system of animals. The killer T-cells, which kill virus infected cells, suicide before they go to throughout body if they attack normal body cells. Likewise, a grammar which generates a negative sample word suicides.

The negative 1 evaluation is performed once in p iteration steps. In a positive evaluation, fitness value is the sum of $\frac{3}{2}|w_t| - d(w_t, w_g)$ for every w_t in the set of positive samples. After $p - 1$ steps with positive evaluation, an iteration step performs negative evaluation in which the fitness value is the sum of $d(w_t, w_g)$ for every w_t in the set of negative samples.

In the selection phase, roulette wheel selection is used in a cell. In addition to this, above mentioned cell death is the selection in the tissue. A simple one point crossover and one point mutation are done between genomes in a cell. The genome migrations from the best cell are the modification using the tissue structure.

We examine 4 types of EP systems:

- Tissue and negative 1 evaluation (tissue NE1).
- Tissue and negative 2 evaluation (or immune-like evaluation, tissue IM).
- Cell-like (using only one cell) and negative 1 evaluation (cell NE1).
- Cell-like and immune-like evaluation (cell IM).

Parameters of the P systems are determined by a number of preliminary experiments. We set total number of genomes to 100. The cell-like P system has 100 genomes while in the tissue P system, one cell has 5 genomes and the whole tissue consists of 20 cells. Thus 100 genomes evolve in both tissue and cell-like P

Table 1. Sample sets of the target languages

	Positive	Negative
Dyck	*aabb*, *abab*, *aaabbb*, *aababb*, *aabbab*, *aaababbb*, *aabbaabb*, *abaabbab*	*baba*, *bbaa*, *a*, *b*, *aaa*, *bbb*, *aabba*, *bbba*
$a^n b^n$	*aabb*, *aaabbb*, *aaaabbbb*, *aaaaabbbbb*	*bbbb*, *bbbaaa*, *abba*, *abab*, *baba*, *abbbaa*, *bbaabb*, *a*, *b*, *aab*, *abb*
ww^R	*aaaa*, *abba*, *baab*, *bbbb*, *aaaaaa*, *bbbbbb*, *abbbba*, *bbaabb*	*aaa*, *bba*, *bbbbb*, *baaa*, *aba*, *aaabbb*, *ab*, *aabb*, *a*, *b*, *bbaa*

systems. In the initialization, a genome is randomly made with having at most 5 rules of length of the right-hand side at most 4. The negative 1 evaluation evaluates using negative samples once in 3 iterations in tissue P systems and once in 10 iterations in cell-like P systems. For Dyck and $a^n b^n$ target languages, 100 iterations get some correct grammars in 20 trials. But the P systems of all types cannot evolve correct grammars in 100 iterations for the target ww^R, then iteration steps 100, 200, 300, 400, and 500 are examined as the termination conditions for this case. The positive and negative sample sets are shown in Table 1.

Table 2. Results from Dyck target.

Type	Correct	Semi-correct	Fake
Tissue IM	4	0	0
Tissue NE1	3	0	0
Cell IM	1	0	0
Cell NE1	1	0	2

If the evolved grammar generates the same language as the target language, the trial of evolution is success and is said to be "correct". If the grammar generates all words in the target language but it generates finitely exceptional words, too, then the trial is said to be "semi-correct". If the language generated by the grammar includes the target language but contains infinitely many other words, then the trial is said to be "fake maximal" or "fake". Using the production rules obtained from the system, correct, semi-correct, and fake are classified manually. The numbers of correct, semi-correct and fake grammars in 20 trials are shown in Tables 2, 3, and 4.

5 Discussions

Most preceding genetic algorithms for grammatical inference use grammar specific operations, e.g., crossover points are selected only from end points of pro-

Table 3. Results from $a^n b^n$ target.

Type	Correct	Semi-correct	Fake
Tissue IM	6	3	0
Tissue NE1	2	4	6
Cell IM	4	3	3
Cell NE1	5	1	4

Table 4. Results from ww^R target.

Type	Results
Tissue IM	1 correct (in 300 steps)
Tissue NE1	1 correct (in 500 steps), 2 fakes (1 in 200 steps, 1 in 400 steps)
Cell IM	1 fake (in 200 steps)
Cell NE1	2 corrects (in 300 steps), 9 fakes (2 in 100 steps, 1 in 200 steps, 3 in 300 steps, 1 in 400 steps, 2 in 500 steps)

duction rules, useless rules (do not generate terminal words or are not generated from the start symbol) are removed from the genome, etc. The EP systems in this paper do not use such grammar specific operators since we want to make it clear whether evolution mechanisms of living organisms evolve context-free grammars or not.

Tissue EP system can be seen as a model of cooperation and competition among cells in a tissue. From Tables 2 and 3, tissue EP system infer grammars generating $a^n b^n$ and Dyck better than cell-like EP systems. The migration mechanism seems to keep variety of genomes and to prevent from falling local minima. Negative selection from immune method prevents fake grammars and, by suppressing fake grammars, infers more correct (at least semi-correct) grammars. From Table 4, the EP systems, especially tissue system, are not good at inferring grammars generating ww^R. Some grammar specific operators might be necessary to infer the language. Further experiments using grammar specific operators will make this clear.

References

1. Aman, B., Ciobanu, G.: Adaptive P systems. In: Hinze, T., Rozenberg, G., Salomaa, A., Zandron, C. (eds.) CMC 2018. LNCS, vol. 11399, pp. 57–72. Springer, Cham (2019). https://doi.org/10.1007/978-3-030-12797-8_5
2. Gold, E.M.: Learning identification in the limit. Inf. Control **10**, 447–474 (1967)
3. Andreu-Guzmán, J.A., Valencia-Cabrera, L.: A novel solution for GCP based on an OLMS membrane algorithm with dynamic operators. J. Membr. Comput. **2**, 1–13 (2020)

4. Lankhorst, M.M.: A genetic algorithm for grammatical inference. In: Gentzsch, W., Harms, U. (eds.) HPCN-Europe 1994. LNCS, vol. 796, pp. 418–419. Springer, Heidelberg (1994). https://doi.org/10.1007/BFb0020409
5. Črepinšek, M., Mernik, M., Bryant, B.R., Javed, F., Sprague, A.: Inferring context-free grammars for domain-specific languages. Electron. Notes Theor. Comput. Sci. **141**, 99–116 (2005)
6. Nishida, T.Y.: An application of P-systems: a new algorithm for NP-complete optimization problems. In: Proceedings of the 8th World Multi-conference on Systems, Cybernetics and Informatics, pp. 109–112 (2004)
7. Nishida, T.Y.: Membrane algorithm with Brownian subalgorithm and genetic subalgorithm. Int. J. Found. Comput. Sci. **18**, 1353–1360 (2007)
8. Nishida, T.Y.: Evolutionary P systems (extended abstract). In: Pre-proceedings of CMC 2019, pp. 183–187 (2019)
9. Nishida, T.Y., Shiotani, T., Takahashi, Y.: Membrane Algorithms, pp. 529–552. Oxford University Press (2010)
10. Pandey, H.M.: Genetic algorithm for grammar induction and rules verification through a PDA simulator. IAES Int. J. Artif. Intell. **6**, 100–111 (2017)
11. Păun, G.: Membrane Computing. Springer, Heidelberg (2002). https://doi.org/10.1007/978-3-642-56196-2
12. Păun, G., Rozenberg, G., Salomaa, A. (eds.): Handbook of Membrane Computing. Oxford University Press, Oxford (2010)
13. Stevenson, A., Cordy, J.R.: Grammatical inference in software engineering: an overview of the state of the art. In: Czarnecki, K., Hedin, G. (eds.) SLE 2012. LNCS, vol. 7745, pp. 204–223. Springer, Heidelberg (2013). https://doi.org/10.1007/978-3-642-36089-3_12
14. Wyard, P.: Representational issues for context free grammar induction using genetic algorithms. In: Carrasco, R.C., Oncina, J. (eds.) ICGI 1994. LNCS, vol. 862, pp. 222–235. Springer, Heidelberg (1994). https://doi.org/10.1007/3-540-58473-0_151
15. Zhang, G., Gheorghe, M., Pan, L., Prez-Jiménez, M.J.: Evolutionary membrane computing: a comprehensive survey and new results. Inf. Sci. **279**, 528–551 (2014)
16. Zhang, G., Gheorghe, M., Wu, C.: A quantum-inspired evolutionary algorithm based on P systems for knapsack problem. Fund. Inform. **87**, 93–116 (2008)
17. Zhang, G., Liu, C., Rong, H.: Analyzing radar emitter signals with membrane algorithms. Math. Comput. Model. **52**, 1997–2010 (2010)
18. Zhang, G., Pérez-Jiménez, M.J., Gheorghe, M.: Membrane algorithms. In: Real-Life Applications with Membrane Computing. ECC, vol. 25, pp. 33–115. Springer, Cham (2017). https://doi.org/10.1007/978-3-319-55989-6_3

Partial Array Token Petri Net and P System

K. Sasikala[1], F. Sweety[1], T. Kalyani[2(✉)], and D. G. Thomas[3]

[1] Department of Mathematics, St. Joseph's college of Engineering, Chennai 600119, India
[2] Department of Mathematics, St. Joseph's Institute of Technology, Chennai 600119, India
[3] Department of Applied Mathematics, Saveetha School of Engineering, SIMATS, Chennai 602105, India

Abstract. The innovative model of partial array languages generated by basic puzzle partial array grammars is available in the literature. Here we define Partial array Token Petri Net Structure (PATPNS) to generate partial array languages. Further we introduce Partial Array Token Petri Net P System (PATPNPS) to generate partial array languages and compared with basic puzzle partial array grammars for generative power. PATPNS is also compared with local and recognizable partial array languages.

Keywords: Partial array · Basic puzzle partial array grammar · Partial Array Token Petri Net

1 Introduction

In the context of a syntactic approach to pattern recognition, there have been several studies in the last few decades on theoretical models for generating or recognizing two-dimensional objects, pictures and picture languages [2]. Picture languages generated by several array grammars, matrix grammars have been advocated since the seventies and they have been applied in practical problems such as character recognition, pattern recognition, kolam patterns and tiling systems.

Petri nets are the models in mathematics proposed to model dynamic systems. To simulate the activity of the dynamic system tokens are used, represented by black dots. The tokens move when the transition fires. Array token Petri nets [3–7, 10] are proposed to generate array languages. The arrays over an alphabet are used as tokens over an alphabet not the black dots. The transitions are associated with catenation rules. Firing of transitions helps to catenate the arrays to build bigger arrays.

The class of grammars with array rewriting methods is a dynamic device to describe picture languages. The picture languages produced by such array

R. Freund et al. (Eds.): CMC 2020, LNCS 12687, pp. 135–152, 2021.
https://doi.org/10.1007/978-3-030-77102-7_8

grammars, matrix grammars have been applied in practical problems. Nivat et al. [8] proposed puzzle grammars to generate two-dimensional picture languages.

Partial words were introduced by Berstel and Boasson [1]. Later on, Partial array languages were introduced in [14] and the combinatorial properties of partial array languages were studied in [15]. Gh. Paun [9] introduced a computational model, called P system. The notion of P system related to arrays can be seen in [12]. Partial array grammars and partial array rewriting - P system were introduced in [11]. We have proposed Basic Puzzle Partial Array Grammar (BPPAG) to generate Partial array languages and studied the generative capacity of the resulting partial array P system with BPPAG [13]. Motivated by these studies, in this paper, we introduce Partial Array Token Petri Net Structure (PATPNS) and Partial Array Token Petri Net P System (PATPNPS) to generate partial array languages and we examine the generative capacity of both systems and give some comparison results. PATPNS is compared with local and recognizable partial array languages and we have proved that PATPNS has more generative power.

2 Preliminaries

The basic concepts and definitions of Partial Word, Partial Array, Basic Puzzle Partial Array Grammar and Petri Net are given here with examples.

Definition 1. *[1] A partial word u of length n over Σ, is a partial function $u : N \rightarrow \Sigma$. For $1 \leq i \leq n$, if $u(i)$ is defined, then we say that i belongs to the domain of u (denoted by $i \in D(u)$); Otherwise, we say that i belongs to the set of holes of u (denoted by $i \in H(u)$). A word over Σ is a partial word over Σ with an empty set of holes. $H(u)$ is the set of positions in which the 'do not know' symbol '$\diamondsuit$' appears in u.*

Definition 2. *[1] If u is a partial word of length n over Σ, then the companion of u (denoted by $u_\diamondsuit$) is the total function $u_\diamondsuit : N \rightarrow \Sigma \cup \{\diamondsuit\}$ defined by*

$$u_\diamondsuit(i) = \begin{cases} u(i), i \in D(u); \\ \diamondsuit, \quad \textit{otherwise}. \end{cases} \quad \textit{where} \quad \diamondsuit \notin \Sigma.$$

The symbol '$\diamondsuit$' is viewed as a 'do not know' symbol and not as a 'do not care' symbol as in pattern matching.

Definition 3. *[14] A partial array A of size $m \times n$ over Σ is a partial function $A : Z_+^2 \rightarrow \Sigma$, where Z is the set of all positive integers. For $1 \leq i \leq m$, $1 \leq j \leq n$, if $A(i,j)$ is defined then we say that (i,j) belongs to the domain of A (denoted by $(i,j) \in D(A)$); Otherwise, we say that (i,j) belongs to the set of holes of A (denoted by $(i,j) \in H(A)$). An array over Σ is partial array over Σ with an empty set of holes. $H(A)$ is the set of positions in which the 'do not know' symbol '$\diamondsuit$' appears in A.*

Definition 4. *[14] If A is a partial array of size $m \times n$ over Σ, then the companion of A (denoted by $A_\diamondsuit$) is the total function $A_\diamondsuit : Z_+^2 \rightarrow \Sigma \cup \{\diamondsuit\}$ defined by*

$$A_\diamondsuit(i,j) = \begin{cases} A(i,j), (i,j) \in D(A); \\ \diamondsuit, \quad \text{otherwise.} \end{cases} \quad \text{where} \quad \diamondsuit \notin \Sigma.$$

Example 1. [14] The Partial array, $A_\diamondsuit = \begin{pmatrix} a & b & a \\ \diamondsuit & b & a \\ a & \diamondsuit & b \end{pmatrix}$ is a companion of a partial array A of size $(3,3)$ where $D(A) = \{(1,1),(1,2),(1,3),(2,2),(2,3),(3,1),(3,3)\}$ and $H(A) = \{(2,1),(3,2)\}$

Definition 5. *[14] If A and B are two partial arrays of equal size, then A is contained in B, denoted by $A \subset B$ if $D(A) \subseteq D(B)$ and $A(i,j) = B(i,j)$ for all $(i,j) \in D(A)$. The partial arrays A and B are said to be compatible, denoted by $A \uparrow B$ if there exists a partial array C such that $A \subset C$ and $B \subset C$.*
$A_\diamondsuit = \begin{pmatrix} a & b & a \\ \diamondsuit & b & a \\ a & \diamondsuit & b \end{pmatrix}$ and $B_\diamondsuit = \begin{pmatrix} \diamondsuit & b & \diamondsuit \\ a & b & a \\ a & \diamondsuit & b \end{pmatrix}$ are the companions of two partial arrays A and B that are compatible.

The set of all partial arrays over Σ is denoted by Σ_p^{**}, where $\Sigma_p = \Sigma \cup \{\diamondsuit\}$. We denote the empty array with no symbols by Λ and $\Sigma_p^{++} = \Sigma_p^{**} - \{\Lambda\}$. The set of all partial arrays over Σ of size (k,r), $k \leq m, r \leq n$ is denoted by $\Sigma_p^{(k,r)}$.

Definition 6. *[13] The structure of a Basic Puzzle Partial Array Grammar (BPPAG) is $BPG_p = (A, B \cup \{\diamondsuit\}, P, S)$ where A is a finite non empty set of non terminal symbols and B is a finite non empty set of terminal symbols. '$\diamondsuit$' is a 'do not know' symbol, where $\diamondsuit \notin A \cup B$, $S \in A$ is the axiom pattern and P is a set of rules of the following forms:*

(*i*) $X \rightarrow \textcircled{x}Y$ (*ii*) $X \rightarrow \textcircled{\diamondsuit}Y$ (*iii*) $X \rightarrow \textcircled{Y}x$

(*iv*) $X \rightarrow \textcircled{Y}\diamondsuit$ (*v*) $X \rightarrow Y\textcircled{x}$ (*vi*) $X \rightarrow Y\textcircled{\diamondsuit}$

(*vii*) $X \rightarrow x\textcircled{Y}$ (*viii*) $X \rightarrow \diamondsuit\textcircled{Y}$ (*ix*) $X \rightarrow \begin{array}{c}\textcircled{x}\\ Y\end{array}$

(*x*) $X \rightarrow \begin{array}{c}\textcircled{\diamondsuit}\\ Y\end{array}$ (*xi*) $X \rightarrow \begin{array}{c}\textcircled{Y}\\ x\end{array}$ (*xii*) $X \rightarrow \begin{array}{c}\textcircled{Y}\\ \diamondsuit\end{array}$

(*xiii*) $X \rightarrow \begin{array}{c}Y\\ \textcircled{x}\end{array}$ (*xiv*) $X \rightarrow \begin{array}{c}Y\\ \textcircled{\diamondsuit}\end{array}$ (*xv*) $X \rightarrow \begin{array}{c}x\\ \textcircled{Y}\end{array}$

(*xvi*) $X \rightarrow \begin{array}{c}\diamondsuit\\ \textcircled{Y}\end{array}$ (*xvii*) $X \rightarrow \textcircled{x}$ (*xviii*) $X \rightarrow \textcircled{\diamondsuit}$

where $X, Y \in A$ and $x, y \in B$.

While processing the derivations in the production rule $X \rightarrow \textcircled{x}Y$, the nonterminal X is replaced by the right-hand member whose left-hand side is X.

The replacement is possible only if the noncircled symbol of the production rule consists of a blank symbol. The blank symbol is represented by the letter '#', which is an unoccupied place where any symbol can be occupied as per the derivation. The language generated by BPPAG is denoted by $\mathcal{L}(BPPAG)$.

Example 2. [13] Consider a BPPAG $BPG_{p_1} = (A, B \cup \{\diamondsuit\}, P, S)$ where $A = \{X, Q, R, S_1, T, U, V\}$, $B = \{z\}$, $S = X$ and P consists of the following rules:

$(i)\ \ X \rightarrow \textcircled{z}Q$ $\qquad (ii)\ \ Q \rightarrow \textcircled{z}Q$ $\qquad (iii)\ \ Q \rightarrow \begin{array}{c} R \\ \textcircled{z} \end{array}$

$(iv)\ \ R \rightarrow S_1\textcircled{z}$ $\qquad (v)\ \ S_1 \rightarrow \textcircled{z}$ $\qquad (vi)\ \ S_1 \rightarrow S_1\textcircled{$\diamondsuit$}$

$(vii)\ \ S_1 \rightarrow \begin{array}{c} T \\ \textcircled{z} \end{array}$ $\qquad (viii)\ \ T \rightarrow \textcircled{z}T$ $\qquad (ix)\ \ T \rightarrow \textcircled{z}$

$(x)\ \ T \rightarrow \textcircled{$\diamondsuit$}T$ $\qquad (xi)\ \ T \rightarrow \begin{array}{c} U \\ \textcircled{z} \end{array}$ $\qquad (xii)\ \ U \rightarrow U\textcircled{z}$

$(xiii)\ \ U \rightarrow \textcircled{z}$ $\qquad (xiv)\ \ T \rightarrow \begin{array}{c} R \\ \textcircled{z} \end{array}$

This grammar generates square partial arrays of size $(m \times m, m \geq 2)$ with $(m - 2 \times m - 2, m \geq 2)$ square partial array in the center consisting of only $\{\diamondsuit\}$ symbol bounded by the terminal alphabet 'z' on the boundary of the square, for $m = 2$, the grammar generates 2×2 square array $\begin{array}{cc} z & z \\ z & z \end{array}$.

The first three members of the language are given below:

$$\begin{array}{cc} z & z \\ z & z \end{array} \qquad \begin{array}{ccc} z & z & z \\ z & \diamondsuit & z \\ z & z & z \end{array} \qquad \begin{array}{cccc} z & z & z & z \\ z & \diamondsuit & \diamondsuit & z \\ z & \diamondsuit & \diamondsuit & z \\ z & z & z & z \end{array} \qquad \cdots$$

A Petri Net [10] is an abstract formal model of information flow. Petri nets have been used for analyzing systems that are concurrent, asynchronous, distributed, parallel, non-deterministic and/or stochastic. Tokens are used in Petri Nets to simulate dynamic and concurrent activities of the system. A language can be associated with the execution of a Petri Net. By defining a labeling function for transitions over an alphabet, the set of all firing sequences, starting from a specific initial marking leading to a finite set of terminal markings, generates a language over the alphabet. Petri Net structure to generate rectangular arrays are found in [3–5]. The two models have different firing rules and catenation rules. In [6], Column Row Catenation Petri Net Structure (CRCPNS) has been defined. Several input places having different arrays is associated with a catenation rules label. The label of the transition decides the order in which the arrays are joined (column wise or row wise) provided the condition for catenation is satisfied. In CRCPNS a transition with a catenation rule as label and different arrays in the input places is enabled to fire. In ATPNS [15] the catenation rule involves an array language. All the input places of the transition with a catenation rule as label, should have the same array as token, for the transition to be enabled. The size of the array language to be joined to the array in the input place, depends on the size of the array in the input place.

Definition 7. *[5] A Petri Net structure is a four tuple $C = (P, T, I, O)$ where $P = \{p_1, p_2, \ldots, p_n\}$ is a finite set of places, $n > 0$, $T = \{t_1, t_2, \ldots, t_m\}$ is a finite set of transitions, $m > 0$, $P \cap T = \phi$, $I : T \to P^\infty$ is the input function from transitions to bags of places and $O : T \to P^\infty$ is the output function from transitions to bags of places, where P^∞ is the bags of places.*

Definition 8. *[5] A Petri Net marking is an assignment of tokens to the places of Petri Net. The tokens are used to define the execution of a Petri Net. The number and position of tokens may change during the execution of a Petri Net, arrays over an alphabet are used as tokens.*

3 Partial Array Token Petri Net Structure

In this section, we define Partial Array Token Petri Net Structure (PATPNS) with an example and compare it with basic puzzle partial array languages.

Definition 9. *If $C = (P, T, I, O)$ is a Petri Net structure with partial arrays over $(\Sigma \cup \{\Diamond\})^{**}$ as initial markings. $\mu_0 : P \to (\Sigma \cup \{\Diamond\})^{**}$ label of at least one transition being catenation rule and a finite set of final places $F \subset P$, then the Petri net structure C is defined as a Partial Array Token Petri Net Structure (PATPNS).*

Definition 10. *If C is a PATPNS, then the Partial array language generated by the Petri Net C is defined as*

$$PL(C) = \{A_\Diamond \in (\Sigma \cup \{\Diamond\})^{**} \ / \ A_\Diamond \ is \ in \ p \ for \ some \ p \ in \ F\}$$

with partial arrays over $(\Sigma \cup \{\Diamond\})^{**}$ in some places as initial marking when all possible sequences of transitions are fired. The set of all partial arrays collected in the final places F is called the partial array language generated by C. Let $\mathcal{L}(PATPNS) = \{PL(C)/C \text{ is a } PATPNS\}$.

$(\Sigma \cup \{\Diamond\})^{**}$ denotes the partial arrays made up of elements of $\Sigma \cup \{\Diamond\}$. If A and B are two partial arrays having same number of rows then $A \obar B$ is the column wise catenation of A and B. If two partial arrays have the same number of columns then $A \ominus B$ is the row wise catenation of A and B. $(x)^n$ denotes a horizontal sequence of n 'x' and $(x)_n$ denotes a vertical sequence of n 'x' where $x \in (\Sigma \cup \{\Diamond\})^{**}$, $(x)^{n+1} = (x)^n \obar x$ and $(x)_{n+1} = (x)_n \ominus x$.

The Petri Net model defined here has places and transitions connected by directed arcs. Rectangular partial arrays over an alphabet are taken as tokens to be distributed in places. Variation in firing rules and labels of the transition are listed out below.

Firing Rules in PATPNS

We define three different types of enabled transition in PATPNS. The pre and post condition for firing the transition in all the three cases are given below:

1. When all the input places of t_1 (without label) have the same partial array as token.
   - Each input place should have at least the required number of partial arrays.
   - Firing t_1 removes partial array from all the input places and moves the partial array to all its output places.

   The graph in Fig. 1 shows the position of the partial array before the transition fires and Fig. 2 shows the position of the partial array after transition t_1 fires.

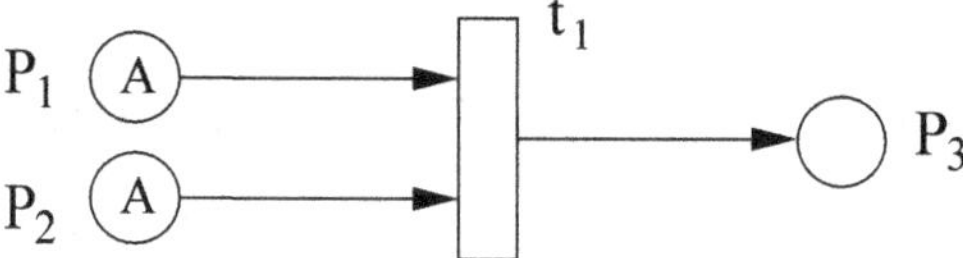

Fig. 1. Position of partial array before firing

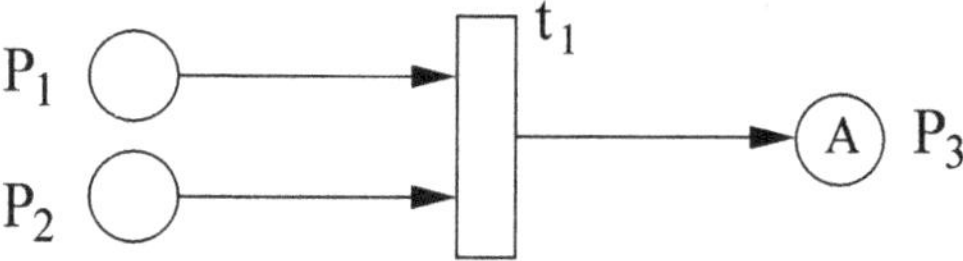

Fig. 2. Position of partial array after firing

2. When all the input places of t_1 have different partial arrays as token
   - The label of t_1 designates one of its input places.
   - The designated input place has sufficient number of partial arrays as tokens.
   - Firing t_1 removes partial array from all the input places and moves the partial array from the designated input place to all its output places.

   The graph in Fig. 3 shows the position of the partial array before the transition fires and Fig. 4 shows the position of the partial array after transition t_1 fires. Since the designated place is P_1, the partial array in P_1 is moved to the output place.
3. When all the input places of t_1 (with catenation rule as label) have the same partial array as token
   - Each input place should have at least the required number of partial arrays.

- The condition for catenation should be satisfied.
- The designated input place has sufficient number of partial arrays as tokens.
- Firing t_1 removes partial array from all the input places P and the catenation is carried out in all its output places.

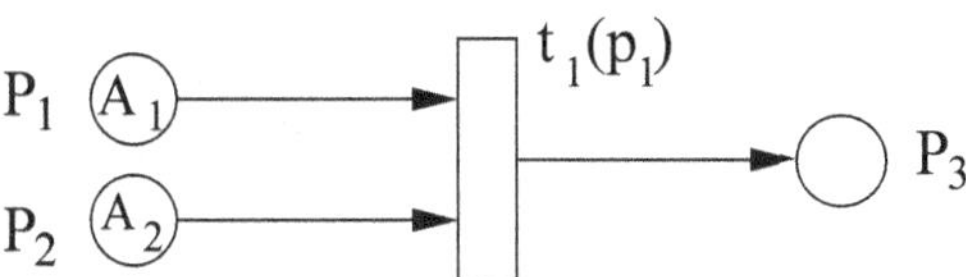

Fig. 3. Transition with label before firing

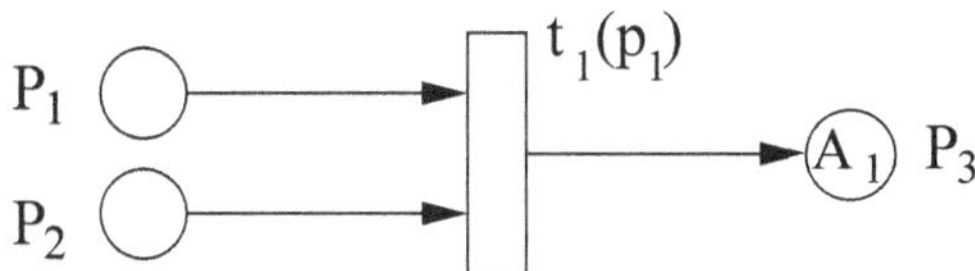

Fig. 4. Transition with label after firing

Catenation Rule as Label for Transitions

Column catenation rule is in the form $A \textcircled{|} B$. Here the partial array A denotes the $m \times n$ partial array in the input place of the transition. B is a partial array whose number of rows will depend on 'm', the number of rows of A. The number of columns of B is fixed. For example $A \textcircled{|} (x \quad x)_m$ adds two columns of x after the last column of the partial array A which is in the input place. But $(x \quad x)_m \textcircled{|} A$ would add two columns of x before the first column of A. 'm' always denote the number of rows of the input partial array A. Row catenation rule is in the form $A \textcircled{-} B$. Here again the partial array A denotes the $m \times n$ partial array in the input place of the transition. B is a partial array whose number of columns will depend on 'n', the number of columns of A. The number of rows of B is always fixed. For example $A \textcircled{-} \begin{bmatrix} x \\ x \end{bmatrix}^n$ adds two rows of x after the last row of the array A which is in the input place. But $\begin{bmatrix} x \\ x \end{bmatrix}^n \textcircled{-} A$ would add two rows of x before the first row of the partial array A. 'n' always denotes the number of columns of the input partial array A.

An example to explain row catenation rule is given below. The position of the partial array before the transition fires is shown in Fig. 5 and Fig. 6 shows

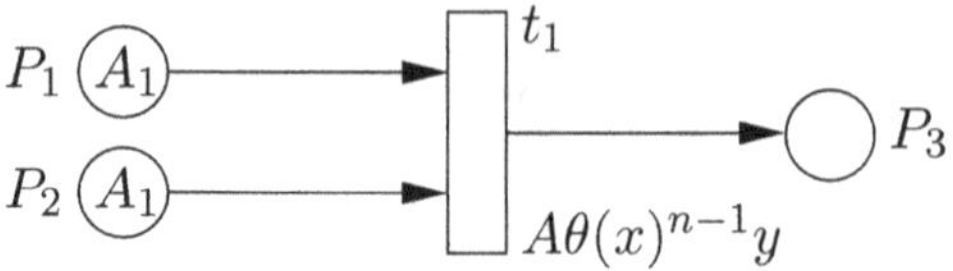

Fig. 5. Transition with catenation rule before firing

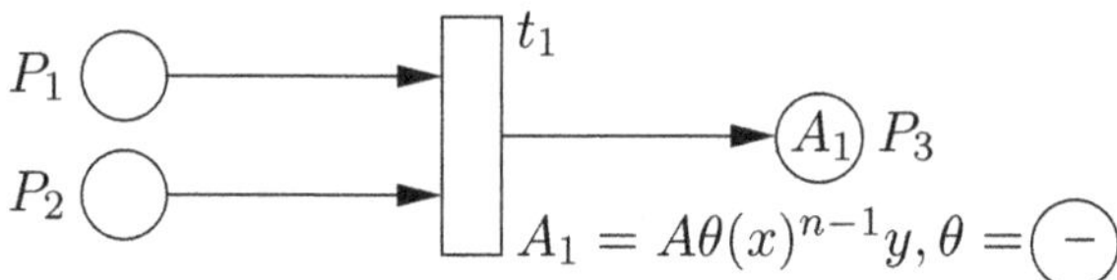

Fig. 6. Transition with catenation rule after firing

the position of the partial array after transition t_1 fires. Since the catenation rule is associated with the transition, catenation takes place in P_3.

In $A_\diamondsuit = \begin{array}{ccc} a & a & a \\ a & \diamondsuit & a \\ a & a & a \end{array}$, the number of columns of A is 3, $n-1$ is 2, firing t_1 adds the row $x\ x\ y$ as the last row. Hence $A_{1\diamondsuit} = \begin{array}{ccc} a & a & a \\ a & \diamondsuit & a \\ a & a & a \\ x & x & y \end{array}$

Example 3. Let $\Sigma = \{a\}$, $F = P_1$, where $S_\diamondsuit = \begin{array}{ccc} a & a & a \\ a & \diamondsuit & a \\ a & a & a \end{array}$, $Q_1 = (\diamondsuit)_m, Q_2 = (\diamondsuit)^n$ $Q_3 = (a)_m$, $Q_4 = (a)^n$

S is the initial partial array placed in P_1. The PATPNS is shown in Fig. 7. Derivations in PATPNS is given in the following tabular column.

Input place	Transition	Output place
S	$A \textcircled{\vert} Q_1$	$\begin{array}{cccc} a & a & a & \diamondsuit \\ a & \diamondsuit & a & \diamondsuit \\ a & a & a & \diamondsuit \end{array}$
$\begin{array}{cccc} a & a & a & \diamondsuit \\ a & \diamondsuit & a & \diamondsuit \\ a & a & a & \diamondsuit \end{array}$	$Q_1 \textcircled{\vert} A$	$\begin{array}{ccccc} \diamondsuit & a & a & a & \diamondsuit \\ \diamondsuit & a & \diamondsuit & a & \diamondsuit \\ \diamondsuit & a & a & a & \diamondsuit \end{array}$
$\begin{array}{ccccc} \diamondsuit & a & a & a & \diamondsuit \\ \diamondsuit & a & \diamondsuit & a & \diamondsuit \\ \diamondsuit & a & a & a & \diamondsuit \end{array}$	$A \ominus Q_2$	$\begin{array}{ccccc} \diamondsuit & a & a & a & \diamondsuit \\ \diamondsuit & a & \diamondsuit & a & \diamondsuit \\ \diamondsuit & a & a & a & \diamondsuit \\ \diamondsuit & \diamondsuit & \diamondsuit & \diamondsuit & \diamondsuit \end{array}$

Input place	Transition	Output place
$\diamondsuit$ a a a $\diamondsuit$ $\diamondsuit$ a $\diamondsuit$ a $\diamondsuit$ $\diamondsuit$ a a a $\diamondsuit$ $\diamondsuit$ $\diamondsuit$ $\diamondsuit$ $\diamondsuit$ $\diamondsuit$	Q_2 (−) A	$\diamondsuit$ $\diamondsuit$ $\diamondsuit$ $\diamondsuit$ $\diamondsuit$ $\diamondsuit$ a a a $\diamondsuit$ $\diamondsuit$ a $\diamondsuit$ a $\diamondsuit$ $\diamondsuit$ a a a $\diamondsuit$ $\diamondsuit$ $\diamondsuit$ $\diamondsuit$ $\diamondsuit$ $\diamondsuit$
$\diamondsuit$ $\diamondsuit$ $\diamondsuit$ $\diamondsuit$ $\diamondsuit$ $\diamondsuit$ a a a $\diamondsuit$ $\diamondsuit$ a $\diamondsuit$ a $\diamondsuit$ $\diamondsuit$ a a a $\diamondsuit$ $\diamondsuit$ $\diamondsuit$ $\diamondsuit$ $\diamondsuit$ $\diamondsuit$	A (|) Q_3	$\diamondsuit$ $\diamondsuit$ $\diamondsuit$ $\diamondsuit$ $\diamondsuit$ a $\diamondsuit$ a a a $\diamondsuit$ a $\diamondsuit$ a $\diamondsuit$ a $\diamondsuit$ a $\diamondsuit$ a a a $\diamondsuit$ a $\diamondsuit$ $\diamondsuit$ $\diamondsuit$ $\diamondsuit$ $\diamondsuit$ a
$\diamondsuit$ $\diamondsuit$ $\diamondsuit$ $\diamondsuit$ $\diamondsuit$ a $\diamondsuit$ a a a $\diamondsuit$ a $\diamondsuit$ a $\diamondsuit$ a $\diamondsuit$ a $\diamondsuit$ a a a $\diamondsuit$ a $\diamondsuit$ $\diamondsuit$ $\diamondsuit$ $\diamondsuit$ $\diamondsuit$ a	Q_3 (|) A	a $\diamondsuit$ $\diamondsuit$ $\diamondsuit$ $\diamondsuit$ $\diamondsuit$ a a $\diamondsuit$ a a a $\diamondsuit$ a a $\diamondsuit$ a $\diamondsuit$ a $\diamondsuit$ a a $\diamondsuit$ a a a $\diamondsuit$ a a $\diamondsuit$ $\diamondsuit$ $\diamondsuit$ $\diamondsuit$ $\diamondsuit$ a
a $\diamondsuit$ $\diamondsuit$ $\diamondsuit$ $\diamondsuit$ $\diamondsuit$ a a $\diamondsuit$ a a a $\diamondsuit$ a a $\diamondsuit$ a $\diamondsuit$ a $\diamondsuit$ a a $\diamondsuit$ a a a $\diamondsuit$ a a $\diamondsuit$ $\diamondsuit$ $\diamondsuit$ $\diamondsuit$ $\diamondsuit$ a	A (−) Q_4	a $\diamondsuit$ $\diamondsuit$ $\diamondsuit$ $\diamondsuit$ $\diamondsuit$ a a $\diamondsuit$ a a a $\diamondsuit$ a a $\diamondsuit$ a $\diamondsuit$ a $\diamondsuit$ a a $\diamondsuit$ a a a $\diamondsuit$ a a $\diamondsuit$ $\diamondsuit$ $\diamondsuit$ $\diamondsuit$ $\diamondsuit$ a a a a a a a a
a $\diamondsuit$ $\diamondsuit$ $\diamondsuit$ $\diamondsuit$ $\diamondsuit$ a a $\diamondsuit$ a a a $\diamondsuit$ a a $\diamondsuit$ a $\diamondsuit$ a $\diamondsuit$ a a $\diamondsuit$ a a a $\diamondsuit$ a a $\diamondsuit$ $\diamondsuit$ $\diamondsuit$ $\diamondsuit$ $\diamondsuit$ a a a a a a a a	Q_4 (−) A	a a a a a a a a $\diamondsuit$ $\diamondsuit$ $\diamondsuit$ $\diamondsuit$ $\diamondsuit$ a a $\diamondsuit$ a a a $\diamondsuit$ a a $\diamondsuit$ a $\diamondsuit$ a $\diamondsuit$ a a $\diamondsuit$ a a a $\diamondsuit$ a a $\diamondsuit$ $\diamondsuit$ $\diamondsuit$ $\diamondsuit$ $\diamondsuit$ a a a a a a a a

The firing of sequence $(t_1 t_2 t_3 t_4 t_5 t_6 t_7 t_8)^k$, $k \geq 0$ puts a square partial arrays of size $4k + 3$ in P_1, where the boundaries of the squares are alternatively $\diamondsuit$'s and a's. The partial array language generated by the PATPNS is a square partial array of size $4k+3$, $k \geq 0$ where the boundaries are alternatively $\diamondsuit$'s on the odd numbered boundaries and a's on the even numbered boundaries.

Theorem 1. *The family of languages generated by PATPNS is properly contained in the family of languages generated by Basic Puzzle Partial Array Grammars.*

Proof. The row catenation in PATPNS can be handled by the following Basic Puzzle Partial Array Grammar rules:

$(i)\quad X \to \begin{array}{c} \textcircled{x} \\ Y \end{array} \qquad (ii)\quad X \to \begin{array}{c} \textcircled{\diamondsuit} \\ Y \end{array} \qquad (iii)\quad X \to \begin{array}{c} \textcircled{Y} \\ \diamondsuit \end{array} \qquad (iv)\quad X \to \begin{array}{c} \textcircled{Y} \\ x \end{array}$

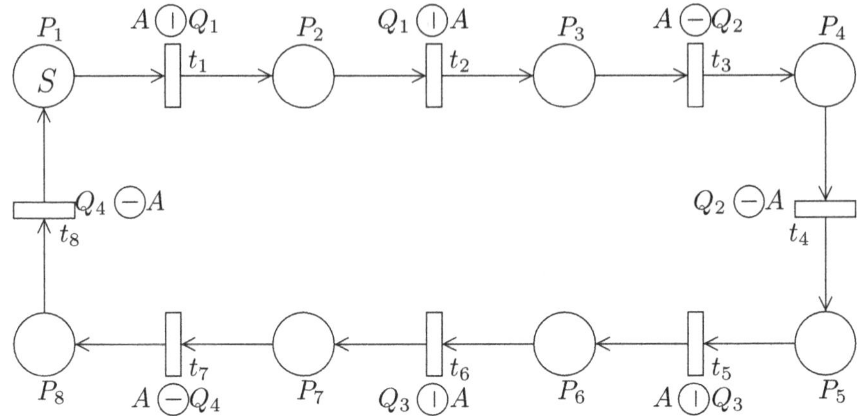

Fig. 7. PATPNS generating square partial arrays of size $4k+3$, $k \geq 0$

$(v)\quad X \rightarrow \begin{array}{c} Y \\ \textcircled{x} \end{array}$ $\quad (vi)\quad X \rightarrow \begin{array}{c} Y \\ \textcircled{\Diamond} \end{array}$ $\quad (vii)\quad X \rightarrow \begin{array}{c} x \\ \textcircled{Y} \end{array}$ $\quad (viii)\quad X \rightarrow \begin{array}{c} \Diamond \\ \textcircled{Y} \end{array}$

$(ix)\quad X \rightarrow \textcircled{x}$ $\quad (x)\quad X \rightarrow \textcircled{\Diamond}$

The column catenation in PATPNS can be handled by the following Basic Puzzle Partial Array Grammar rules:

$(i)\quad X \rightarrow \textcircled{x}\,Y$ $\quad (ii)\quad X \rightarrow \textcircled{\Diamond}\,Y$ $\quad (iii)\quad X \rightarrow \textcircled{Y}\,x$

$(iv)\quad X \rightarrow \textcircled{Y}\,\Diamond$ $\quad (v)\quad X \rightarrow Y\,\textcircled{x}$ $\quad (vi)\quad X \rightarrow Y\,\textcircled{\Diamond}$

$(vii)\quad X \rightarrow \Diamond\,\textcircled{Y}$ $\quad (viii)\quad X \rightarrow x\,\textcircled{Y}$ $\quad (ix)\quad X \rightarrow \textcircled{x}$ $\quad (x)\quad X \rightarrow \textcircled{\Diamond}$

Hence $\mathcal{L}(PATPNS)$ is a subset of $\mathcal{L}(BPPAG)$, this is also evident from the following example.

Consider a partial array language of square partial arrays of size $4k+3$, $k \geq 0$ whose boundaries are alternatively $\Diamond$'s on the odd numbered boundaries and a's on the even numbered boundaries given in Example 3. This partial array language is generated by both systems PATPNS and BPPAG.

Now let us consider a BPPAG generating this partial array language.

$$BPG_{P_2} = (A, B \cup \{\Diamond\}, P, S)$$

where $A = \{X, Q_1, Q_2, Q_3, Q_4, Q_5, Q_6\}$, $B = \{a\}$, $S = X$ and P consists of the following rules:

$(i)\quad X \rightarrow \textcircled{a}Q$ $\quad (ii)\quad Q \rightarrow \textcircled{a}Q$ $\quad (iii)\quad Q \rightarrow \begin{array}{c} Q_1 \\ \textcircled{a} \end{array}$

$(iv)\quad Q_1 \rightarrow Q_2\textcircled{a}$ $\quad (v)\quad Q_2 \rightarrow Q_3\textcircled{\Diamond}$ $\quad (vi)\quad Q_3 \rightarrow \begin{array}{c} Q_4 \\ \textcircled{a} \end{array}$

(vii) $Q_4 \rightarrow \textcircled{a} Q_5$ $(viii)$ $Q_5 \rightarrow \textcircled{a} Q_5$ (ix) $Q_5 \rightarrow \textcircled{a}$

(x) $Q_3 \rightarrow \textcircled{\diamondsuit} Q_3$ (xi) $Q_5 \rightarrow \textcircled{\diamondsuit} Q_6$ (xii) $Q_6 \rightarrow \textcircled{a} Q_6$

$(xiii)$ $Q_6 \rightarrow \textcircled{\diamondsuit} Q$ (xiv) $Q_3 \rightarrow Q_2 \textcircled{a}$ (xv) $Q_3 \rightarrow Q_3 \textcircled{\diamondsuit}$

(xvi) $Q \rightarrow \textcircled{\diamondsuit} Q$

The first member of the language generated is shown below:

$$X \xrightarrow{(i)} \textcircled{a}\, Q \xrightarrow{(ii)} a\, \textcircled{a}\, Q \xrightarrow{(iii)} \begin{matrix} & & Q_1 \\ a & a & \textcircled{a} \end{matrix} \xrightarrow{(iv)} \begin{matrix} & Q_2 & \textcircled{a} \\ a & a & a \end{matrix}$$

$$\xrightarrow{(v)} \begin{matrix} Q_3 & \textcircled{\diamondsuit} & a \\ a & a & a \end{matrix} \xrightarrow{(vi)} \begin{matrix} & Q_4 & \\ a & \diamondsuit & a \\ a & a & a \end{matrix} \xrightarrow{(vii)} \begin{matrix} \textcircled{a} & Q_5 & \\ a & \diamondsuit & a \\ a & a & a \end{matrix} \xrightarrow{(viii)} \begin{matrix} a & \textcircled{a} & Q_5 \\ a & \diamondsuit & a \\ a & a & a \end{matrix} \xrightarrow{(ix)} \begin{matrix} a & a & a \\ a & \diamondsuit & a \\ a & a & a \end{matrix}$$

The partial array language given in Example 2 generated by BPPAG cannot be generated by PATPNS, since the axiom array $\begin{matrix} z & z \\ z & z \end{matrix}$ can only be concatenated either row wise or column wise, but $\diamondsuit$ cannot be inserted, which proves a proper containment.

4 Partial Array Token Petri Net *P* System

In this section, Partial Array Token Petri Net P System (PATPNPS) is introduced and it is compared with PATPNS and BPPAG.

Definition 11. *A Partial Array Token Petri Net P System (PATPNPS)* $\pi = (V, T \cup \{\diamondsuit\}, \#, \mu, F_1, F_2, \ldots, F_m, R_1, R_2, \ldots, R_m, i_0)$ *where* V *is a finite set of column partial arrays and row partial arrays of the form* $Q_1 = (a)^n$ *and* $Q_2 = (a)_m$ *where* $a \in T \cup \{\diamondsuit\}$. T *is a finite set of terminal alphabets. '#' is a blank symbol not in* $T \cup \{\diamondsuit\}$. μ *is a membrane structure with 'm' membranes,* $F_1, F_2, \ldots, F_m$ *are finite set of partial arrays over* $T \cup \{\diamondsuit\}$ *associated with the 'm' regions.* $R_1, R_2, \ldots, R_m$ *are rules associated with the m regions of the form* $\left(\{A \textcircled{-} Q, A \textcircled{|} Q\}, tar, P\right)$, *where* $tar \in \{here, in, out\}$ *and* P *is the output obtained after the catenation rule is applied.*

If the target indication is 'here', the output partial array 'P' remains in the same region, if the target indication is 'in', 'P' goes to the immediate inner region and if the target indication is 'out' it goes to the outer membrane, i_0 is the elementary membrane of μ.

A computation in a partial array token petri net P system is defined in the same way as in array rewriting P system. The set of all partial arrays computed by π with 'm' membranes is denoted by $PATPNPL_m(\pi)$.

Example 4. Consider the Partial Array Token Petri Net P System PATPNPS

$$\pi_1 = (\{Q_1, Q_2, Q_3, Q_4\}, \{a, \diamondsuit\}, \#, [_1[_2[_3]_2]_1, F_{1\diamondsuit}, F_{2\diamondsuit}, F_{3\diamondsuit}, R_1, R_2, R_3, 3)$$

where $Q_1 = (\diamondsuit)_m$, $Q_2 = (\diamondsuit)^n$, $Q_3 = (a)_m$, $Q_4 = (a)^n$, $F_{1\diamondsuit} = \begin{matrix} a & a & a \\ a & \diamondsuit & a \\ a & a & a \end{matrix}$, $F_{2\diamondsuit} = F_{3\diamondsuit} = \phi$,

$$R_1 = \left\{ \begin{matrix} (F_1 \textcircled{|} Q_1, here, P_1), (Q_1 \textcircled{|} P_1, here, P_1), (P_1 \textcircled{-} Q_2, here, P_1), \\ (Q_2 \textcircled{-} P_1, in, P_1), (P_2 \textcircled{|} Q_1, here, P_1) \end{matrix} \right\};$$

$$R_2 = \left\{ \begin{matrix} (P_1 \textcircled{|} Q_3, here, P_2), (Q_3 \textcircled{|} P_2, here, P_2), (P_2 \textcircled{-} Q_4, here, P_2), \\ (Q_4 \textcircled{-} P_2, in, P_2), (Q_4 \textcircled{-} P_2, out, P_2) \end{matrix} \right\}$$

$R_3 = \phi$.

The content of region 1 is $F_{1\diamondsuit} = \begin{matrix} a & a & a \\ a & \diamondsuit & a \\ a & a & a \end{matrix}$. The derivations in PATPNPS is given in the following tabular column:

Region(i)	Content($F_{i\diamondsuit}$)	Rule(R_i)	Target	Resultant partial array (P_i)
1	$a\ a\ a$ $a\ \diamondsuit\ a$ $a\ a\ a$	$F_1 \textcircled{\vert} Q_1$	here	$a\ a\ a\ \diamondsuit$ $a\ \diamondsuit\ a\ \diamondsuit$ $a\ a\ a\ \diamondsuit$
1	$a\ a\ a\ \diamondsuit$ $a\ \diamondsuit\ a\ \diamondsuit$ $a\ a\ a\ \diamondsuit$	$Q_1 \textcircled{\vert} P_1$	here	$\diamondsuit\ a\ a\ a\ \diamondsuit$ $\diamondsuit\ a\ \diamondsuit\ a\ \diamondsuit$ $\diamondsuit\ a\ a\ a\ \diamondsuit$
1	$\diamondsuit\ a\ a\ a\ \diamondsuit$ $\diamondsuit\ a\ \diamondsuit\ a\ \diamondsuit$ $\diamondsuit\ a\ a\ a\ \diamondsuit$	$P_1 \textcircled{-} Q_2$	here	$\diamondsuit\ a\ a\ a\ \diamondsuit$ $\diamondsuit\ a\ \diamondsuit\ a\ \diamondsuit$ $\diamondsuit\ a\ a\ a\ \diamondsuit$ $\diamondsuit\ \diamondsuit\ \diamondsuit\ \diamondsuit\ \diamondsuit$
1	$\diamondsuit\ a\ a\ a\ \diamondsuit$ $\diamondsuit\ a\ \diamondsuit\ a\ \diamondsuit$ $\diamondsuit\ a\ a\ a\ \diamondsuit$ $\diamondsuit\ \diamondsuit\ \diamondsuit\ \diamondsuit\ \diamondsuit$	$Q_2 \textcircled{-} P_1$	in	$\diamondsuit\ \diamondsuit\ \diamondsuit\ \diamondsuit\ \diamondsuit$ $\diamondsuit\ a\ a\ a\ \diamondsuit$ $\diamondsuit\ a\ \diamondsuit\ a\ \diamondsuit$ $\diamondsuit\ a\ a\ a\ \diamondsuit$ $\diamondsuit\ \diamondsuit\ \diamondsuit\ \diamondsuit\ \diamondsuit$
2	$\diamondsuit\ \diamondsuit\ \diamondsuit\ \diamondsuit\ \diamondsuit$ $\diamondsuit\ a\ a\ a\ \diamondsuit$ $\diamondsuit\ a\ \diamondsuit\ a\ \diamondsuit$ $\diamondsuit\ a\ a\ a\ \diamondsuit$ $\diamondsuit\ \diamondsuit\ \diamondsuit\ \diamondsuit\ \diamondsuit$	$P_1 \textcircled{\vert} Q_3$	here	$\diamondsuit\ \diamondsuit\ \diamondsuit\ \diamondsuit\ \diamondsuit\ a$ $\diamondsuit\ a\ a\ a\ \diamondsuit\ a$ $\diamondsuit\ a\ \diamondsuit\ a\ \diamondsuit\ a$ $\diamondsuit\ a\ a\ a\ \diamondsuit\ a$ $\diamondsuit\ \diamondsuit\ \diamondsuit\ \diamondsuit\ \diamondsuit\ a$
2	$\diamondsuit\ \diamondsuit\ \diamondsuit\ \diamondsuit\ \diamondsuit\ a$ $\diamondsuit\ a\ a\ a\ \diamondsuit\ a$ $\diamondsuit\ a\ \diamondsuit\ a\ \diamondsuit\ a$ $\diamondsuit\ a\ a\ a\ \diamondsuit\ a$ $\diamondsuit\ \diamondsuit\ \diamondsuit\ \diamondsuit\ \diamondsuit\ a$	$Q_3 \textcircled{\vert} P_2$	here	$a\ \diamondsuit\ \diamondsuit\ \diamondsuit\ \diamondsuit\ \diamondsuit\ a$ $a\ \diamondsuit\ a\ a\ a\ \diamondsuit\ a$ $a\ \diamondsuit\ a\ \diamondsuit\ a\ \diamondsuit\ a$ $a\ \diamondsuit\ a\ a\ a\ \diamondsuit\ a$ $a\ \diamondsuit\ \diamondsuit\ \diamondsuit\ \diamondsuit\ \diamondsuit\ a$
2	$a\ \diamondsuit\ \diamondsuit\ \diamondsuit\ \diamondsuit\ \diamondsuit\ a$ $a\ \diamondsuit\ a\ a\ a\ \diamondsuit\ a$ $a\ \diamondsuit\ a\ \diamondsuit\ a\ \diamondsuit\ a$ $a\ \diamondsuit\ a\ a\ a\ \diamondsuit\ a$ $a\ \diamondsuit\ \diamondsuit\ \diamondsuit\ \diamondsuit\ \diamondsuit\ a$	$P_2 \textcircled{-} Q_4$	here	$a\ \diamondsuit\ \diamondsuit\ \diamondsuit\ \diamondsuit\ \diamondsuit\ a$ $a\ \diamondsuit\ a\ a\ a\ \diamondsuit\ a$ $a\ \diamondsuit\ a\ \diamondsuit\ a\ \diamondsuit\ a$ $a\ \diamondsuit\ a\ a\ a\ \diamondsuit\ a$ $a\ \diamondsuit\ \diamondsuit\ \diamondsuit\ \diamondsuit\ \diamondsuit\ a$ $a\ a\ a\ a\ a\ a\ a$

Region(i)	Content($F_{i\diamondsuit}$)	Rule(R_i)	Target	Resultant partial array (P_i)
2	$a \diamondsuit \diamondsuit \diamondsuit \diamondsuit \diamondsuit a$ $a \diamondsuit a\ a\ a \diamondsuit a$ $a \diamondsuit a \diamondsuit a \diamondsuit a$ $a \diamondsuit a\ a\ a \diamondsuit a$ $a \diamondsuit \diamondsuit \diamondsuit \diamondsuit \diamondsuit a$ $a\ a\ a\ a\ a\ a\ a$	$Q_4 \ominus P_2$	in or out	$a\ a\ a\ a\ a\ a\ a$ $a \diamondsuit \diamondsuit \diamondsuit \diamondsuit \diamondsuit a$ $a \diamondsuit a\ a\ a \diamondsuit a$ $a \diamondsuit a \diamondsuit a \diamondsuit a$ $a \diamondsuit a\ a\ a \diamondsuit a$ $a \diamondsuit \diamondsuit \diamondsuit \diamondsuit \diamondsuit a$ $a\ a\ a\ a\ a\ a\ a$
3 (If tar=in)	$a\ a\ a\ a\ a\ a\ a$ $a \diamondsuit \diamondsuit \diamondsuit \diamondsuit \diamondsuit a$ $a \diamondsuit a\ a\ a \diamondsuit a$ $a \diamondsuit a \diamondsuit a \diamondsuit a$ $a \diamondsuit a\ a\ a \diamondsuit a$ $a \diamondsuit \diamondsuit \diamondsuit \diamondsuit \diamondsuit a$ $a\ a\ a\ a\ a\ a\ a$	ϕ	-	$a\ a\ a\ a\ a\ a\ a$ $a \diamondsuit \diamondsuit \diamondsuit \diamondsuit \diamondsuit a$ $a \diamondsuit a\ a\ a \diamondsuit a$ $a \diamondsuit a \diamondsuit a \diamondsuit a$ $a \diamondsuit a\ a\ a \diamondsuit a$ $a \diamondsuit \diamondsuit \diamondsuit \diamondsuit \diamondsuit a$ $a\ a\ a\ a\ a\ a\ a$ The output is collected in the elementary membrane
1 (If tar=out)	$a\ a\ a\ a\ a\ a\ a$ $a \diamondsuit \diamondsuit \diamondsuit \diamondsuit \diamondsuit a$ $a \diamondsuit a\ a\ a \diamondsuit a$ $a \diamondsuit a \diamondsuit a \diamondsuit a$ $a \diamondsuit a\ a\ a \diamondsuit a$ $a \diamondsuit \diamondsuit \diamondsuit \diamondsuit \diamondsuit a$ $a\ a\ a\ a\ a\ a\ a$	$P_2 \oplus Q_1$	here	Procedure continues

Thus the partial array language of square partial arrays of size $4k+3$, $k \geq 0$, where the boundaries are alternatively $\diamondsuit$'s on the odd numbered boundaries and a's on the even numbered boundaries is generated by this $PATPNPS$ π_1.

Theorem 2. $\mathcal{L}(PATPNPL_m(\pi)) \cap \mathcal{L}(PATPNS) \neq \phi$.

Proof. The partial array language of square partial arrays of size $4k+3$, $k \geq 0$ is generated by both systems. It is evident from Examples 3 and 4.

Theorem 3. $\mathcal{L}(PATPNPL_m(\pi)) \cap \mathcal{L}(BPPAG) \neq \phi$.

Proof. $\mathcal{L}(PATPNS)$ is a subclass of the family of Basic Puzzle Partial Array Languages by Theorem 1. By Theorem 2, we get that $\mathcal{L}(PATPNPL_m(\pi))$ intersects $\mathcal{L}(PATPNS)$. Thus the two families intersects.

5 Comparative Study with Local and Recognizable Partial Array Languages

In this section, we recall Local and Recognizable Partial Array Languages and compare with PATPNS.

Definition 12. *[14] Let $\Gamma_p = \Gamma \cup \{\diamondsuit\}$ be a finite alphabet. A two dimensional partial array language $PL \subseteq \Gamma_p^{**}$ is local if there exists a finite set θ of tiles over the alphabet $\Gamma_p \cup \{\#\}$ such that $PL = \{A \in \Gamma_p^{**}/B_{2,2}(\hat{A}) \subseteq \theta\}$, where $\hat{A}$ is a partial array surrounded by a special boundary symbol $\# \notin \Gamma$.*

The partial array language PL is local if given such a set θ, we can exactly retrieve the language PL. We call the set θ a representation by tiles for the local language PL and write $PL = L(\theta)$. The family of all local partial array languages is denoted by PAL-LOC.

Example 5. [14] Let $\Gamma_p = \{a, b\} \cup \{\Diamond\}$, and θ be the following set of tiles over $\Gamma_p \cup \{\#\}$.

$$\theta = \left\{ \begin{array}{|c|c|}\hline \# & \# \\ \hline \# & b \\ \hline \end{array}, \begin{array}{|c|c|}\hline \# & \# \\ \hline b & b \\ \hline \end{array}, \begin{array}{|c|c|}\hline \# & \# \\ \hline b & \# \\ \hline \end{array}, \begin{array}{|c|c|}\hline \# & b \\ \hline \# & a \\ \hline \end{array}, \begin{array}{|c|c|}\hline b & b \\ \hline a & \Diamond \\ \hline \end{array}, \begin{array}{|c|c|}\hline b & b \\ \hline \Diamond & b \\ \hline \end{array}, \begin{array}{|c|c|}\hline b & \# \\ \hline b & \# \\ \hline \end{array}, \begin{array}{|c|c|}\hline \# & a \\ \hline \# & a \\ \hline \end{array}, \begin{array}{|c|c|}\hline a & \Diamond \\ \hline a & a \\ \hline \end{array}, \begin{array}{|c|c|}\hline \Diamond & b \\ \hline a & b \\ \hline \end{array}, \right.$$
$$\left. \begin{array}{|c|c|}\hline a & a \\ \hline \# & \# \\ \hline \end{array}, \begin{array}{|c|c|}\hline a & b \\ \hline \# & \# \\ \hline \end{array}, \begin{array}{|c|c|}\hline \# & a \\ \hline \# & \# \\ \hline \end{array}, \begin{array}{|c|c|}\hline b & \# \\ \hline \# & \# \\ \hline \end{array}, \begin{array}{|c|c|}\hline b & b \\ \hline \Diamond & \Diamond \\ \hline \end{array}, \begin{array}{|c|c|}\hline a & \Diamond \\ \hline a & \Diamond \\ \hline \end{array}, \begin{array}{|c|c|}\hline \Diamond & \Diamond \\ \hline \Diamond & \Diamond \\ \hline \end{array}, \begin{array}{|c|c|}\hline \Diamond & b \\ \hline \Diamond & b \\ \hline \end{array}, \begin{array}{|c|c|}\hline \Diamond & \Diamond \\ \hline a & a \\ \hline \end{array} \right\}$$

Then $L(\theta)$ is a partial array language over Γ_p with equal sides of length, the symbols along the top row (the last row) and right most column (the last column are b's, the symbols along the first row (the bottom row) and the left most column (the first column) except the first and last elements of the principal diagonals are a's. The remaining elements of the array are holes.

The first two members of this language are given below:

$$\begin{array}{ccc} b & b & b \\ a & \Diamond & b \\ a & a & b \end{array}, \quad \begin{array}{cccc} b & b & b & b \\ a & \Diamond & \Diamond & b \\ a & \Diamond & \Diamond & b \\ a & a & a & b \end{array}, \quad \cdots$$

Definition 13. *[14] Let Σ be a finite alphabet. A partial array language $PL \subseteq \Sigma_p^{**}$ is called recognizable if there exists a local partial array language PL' over Γ_p and a mapping $\pi : \Gamma_p \to \Sigma_p$ such that $PL = \pi(PL')$, where $\Sigma_p = \Sigma \cup \{\Diamond\}$. The family of all recognizable partial array languages is denoted by PAL-REC.*

Example 6. [14] The set of all partial array languages over one letter alphabet 'a' with all sides of equal length and the symbols along the first row, first column, the last row and last column are holes is not a local partial array language, but it is a recognizable partial array language. This language is obtained from Example 5 by taking a mapping $\pi : \Gamma_p \to \Sigma_p$ where $\Gamma = \{a, b\}$, $\Sigma = \{a\}$ such that $\pi(b) = \pi(a) = \Diamond$ and $\pi(\Diamond) = a$.

Theorem 4. *$PAL - LOC \subsetneq PATPNS$.*

Every local partial language can be easily generated by some PATPNS. Let PL be a partial array language over Γ_p in PAL-LOC with a finite set of tiles θ such that $PL = L(\theta)$.

Consider the PATPNS, $C = (P, T, I, O)$ with partial arrays over Γ_p^{**}, $S_\Diamond = \begin{array}{|c|c|}\hline \# & \# \\ \hline \# & a \\ \hline \end{array}$, $a \in \Gamma_p$. T, the set of all transitions, they can be either row or column catenations.

(i) For all $\begin{array}{|c|c|}\hline \# & \# \\ \hline \# & a \\ \hline \end{array} \in \theta$, where $a \in \Gamma_p$ we define $t_1 = A \textcircled{|} Q_1$, where $Q_1 = \begin{pmatrix} \# \\ b \end{pmatrix}$, $b \in \Gamma_p$.

(ii) For all $\begin{array}{|c|c|}\hline \# & \# \\ \hline a & b \\ \hline\end{array} \in \theta$, $a, b \in \Gamma_p$, we define $t_2 = A \textcircled{|} Q_1$, where $Q_1 = \begin{pmatrix} \# \\ b \end{pmatrix}$, $b \in \Gamma_p$. This transition is repeated till the tile of the form $\begin{array}{|c|c|}\hline \# & \# \\ \hline b & \# \\ \hline\end{array} \in \theta$ is reached and let this process be repeated 'r' $r \geq 0$, no. of times.

(iii) For all $\begin{array}{|c|c|}\hline \# & \# \\ \hline b & \# \\ \hline\end{array} \in \theta$, $b \in \Gamma_p$, we define $t_3 = A \textcircled{|} Q_2$, where $Q_2 = \begin{pmatrix} \# \\ \# \end{pmatrix}$.

(iv) For all $\begin{array}{|c|c|}\hline \# & a \\ \hline \# & b \\ \hline\end{array}, \begin{array}{|c|c|}\hline a & c \\ \hline b & d \\ \hline\end{array}, \begin{array}{|c|c|}\hline c & \# \\ \hline d & \# \\ \hline\end{array} \in \theta$, $a, b, c, d \in \Gamma_p$, we define $t_4 = A \ominus Q_3$, where $Q_3 = (\# \ B \ \#)$, $B \in \Gamma_p^{(1 \times n)}$, where $\Gamma_p^{(1 \times n)}$ is a partial array of size $1 \times n$.

(v) For all tiles of the form $\begin{array}{|c|c|}\hline \# & a \\ \hline \# & \# \\ \hline\end{array}, \begin{array}{|c|c|}\hline a & b \\ \hline \# & \# \\ \hline\end{array}, \begin{array}{|c|c|}\hline b & \# \\ \hline \# & \# \\ \hline\end{array} \in \theta$, $a, b \in \Gamma_p$, we define $t_5 = A \ominus Q_4$, $Q_4 = (\#)_m$, $m \geq 2$. Here, A represents the partial array collected in the output place by the previous transition.

$$P = \{P_1, \underbrace{P_2, P_3, \ldots P_k}_{t_2}, \underbrace{P_{k+1}}_{t_3}, \underbrace{P_{k+2}, P_{k+3}, \ldots P_{k+s+1}}_{t_4}, \underbrace{P_{k+s+2}}_{t_5}\},$$

where P is the set of places, $S_\diamond$ is placed in P_1 initially and then transition t_1 is applied, the resultant partial array is stored in P_2, and then transition t_2 is applied 'r' no. of times. After applying 'r' times the transition t_2, the partial array reaches the place P_k, where $k = r + 2$. After transition t_3 is applied the partial array reaches P_{k+1}. The transition t_4 is repeated 's' $s \geq 0$ number of times until the tile $\begin{array}{|c|c|}\hline \# & a \\ \hline \# & \# \\ \hline\end{array}$ is reached. After this transition, the partial array reaches the place P_{k+s+1}. After transition t_5 is applied the partial array reaches the final place $F = P_{k+s+2}$.

The PATPNS generating the local language PL is given in Fig. 8.

Clearly PATPNS can generate any partial array language in PAL-LOC and hence $PAL - LOC \subseteq PATPNS$.

Now, to prove the proper inclusion, we consider the partial array language given in Example 3, the square partial arrays of size $4k+3$, $k \geq 0$. This language is not local, since the θ set of this language can also generate any array over one letter alphabet 'a' of size $1 \times n$, $n \geq 1$, where θ is given as follows.

$$\theta = \left\{ \begin{array}{|c|c|}\hline \# & \# \\ \hline \# & a \\ \hline\end{array}, \begin{array}{|c|c|}\hline \# & \# \\ \hline a & a \\ \hline\end{array}, \begin{array}{|c|c|}\hline \# & \# \\ \hline a & \# \\ \hline\end{array}, \begin{array}{|c|c|}\hline \# & a \\ \hline \# & a \\ \hline\end{array}, \begin{array}{|c|c|}\hline a & a \\ \hline a & \diamond \\ \hline\end{array}, \begin{array}{|c|c|}\hline a & a \\ \hline \diamond & \diamond \\ \hline\end{array}, \begin{array}{|c|c|}\hline a & a \\ \hline \diamond & a \\ \hline\end{array}, \begin{array}{|c|c|}\hline a & \# \\ \hline a & \# \\ \hline\end{array}, \begin{array}{|c|c|}\hline a & \diamond \\ \hline a & \diamond \\ \hline\end{array}, \begin{array}{|c|c|}\hline \diamond & \diamond \\ \hline \diamond & a \\ \hline\end{array}, \right.$$
$$\left. \begin{array}{|c|c|}\hline \diamond & \diamond \\ \hline a & a \\ \hline\end{array}, \begin{array}{|c|c|}\hline \diamond & \diamond \\ \hline a & \diamond \\ \hline\end{array}, \begin{array}{|c|c|}\hline \diamond & a \\ \hline \diamond & a \\ \hline\end{array}, \begin{array}{|c|c|}\hline a & \diamond \\ \hline a & a \\ \hline\end{array}, \begin{array}{|c|c|}\hline \diamond & a \\ \hline a & a \\ \hline\end{array}, \begin{array}{|c|c|}\hline \diamond & a \\ \hline \diamond & \diamond \\ \hline\end{array}, \begin{array}{|c|c|}\hline a & a \\ \hline \diamond & \diamond \\ \hline\end{array}, \begin{array}{|c|c|}\hline a & \diamond \\ \hline \diamond & \diamond \\ \hline\end{array}, \begin{array}{|c|c|}\hline a & a \\ \hline \# & \# \\ \hline\end{array}, \begin{array}{|c|c|}\hline a & \# \\ \hline \# & \# \\ \hline\end{array} \right\}$$

Hence PAL-LOC is properly contained in PATPNS.

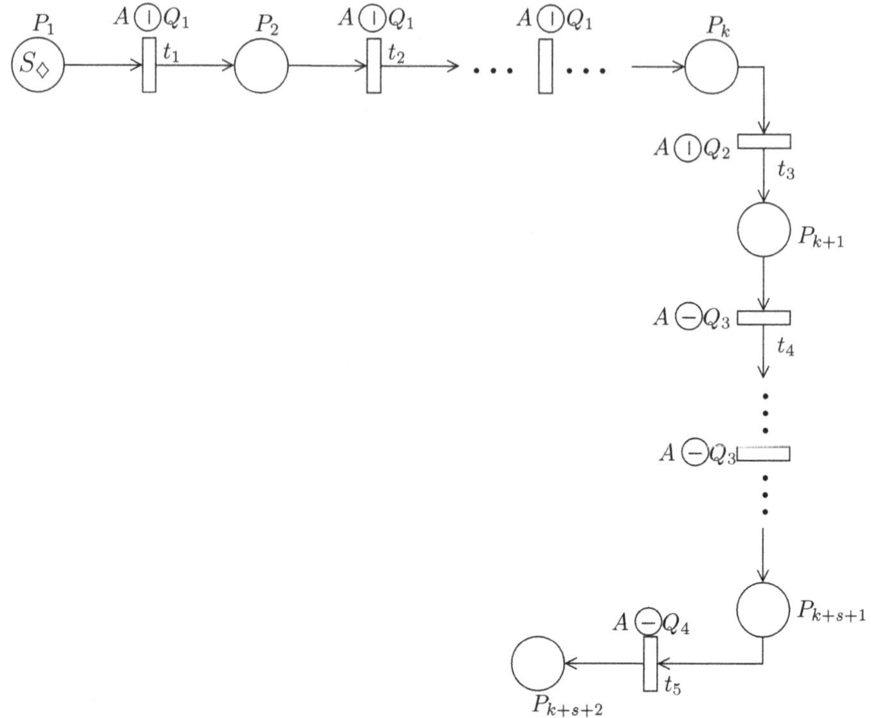

Fig. 8. PATPNS generating any PAL-LOC

Example 7. Consider a PATPNS $C = (P, T, I, O)$, generating a local partial array language given in Example 5.

Let $S_\diamondsuit = \begin{array}{|c|c|} \hline \# & \# \\ \hline \# & b \\ \hline \end{array}$, $Q_1 = \begin{bmatrix} \# \\ b \end{bmatrix}$, $Q_2 = \begin{bmatrix} \# \\ \# \end{bmatrix}$, $Q_3 = [\# \; B \; \#]$, $B \in \Gamma_p^{1 \times n}$, where $B = a\ (\diamondsuit)_r\ b$, $Q_4 = (\#)_m$.

PATPNS generating the 2nd member of the partial array language namely

$$\begin{array}{cccccc} \# & \# & \# & \# & \# & \# \\ \# & b & b & b & b & \# \\ \# & a & \diamondsuit & \diamondsuit & b & \# \\ \# & a & \diamondsuit & \diamondsuit & b & \# \\ \# & a & a & a & b & \# \\ \# & \# & \# & \# & \# & \# \end{array}$$

is given in Fig. 9 as an example.

Theorem 5. *PATPNS is closed under projection.*

We consider a partial array token Petri Net structure $C = (P, T, I, O)$ generating the partial array language PL. Let $\pi : \Gamma_p \rightarrow \Sigma_p$ be a projetion such that $\pi(a) = \alpha$, $a \in \Gamma_p$, $\alpha \in \Sigma_p$. Without loss of generality $\Gamma_p \cap \Sigma_p = \phi$. We can construct a PATPNS $C' = (P', T', I', O')$ such that $PL(C') = PL$, where $T' = \{t'_1, t'_2, \ldots t'_k\}$, $t'_i = \{A' \textcircled{|} Q'_i, A' \ominus Q'_i / A'_{ij} = (\pi(A))_{ij}, (Q'_i)_{rs} = (\pi(Q_i))_{rs}\}$,

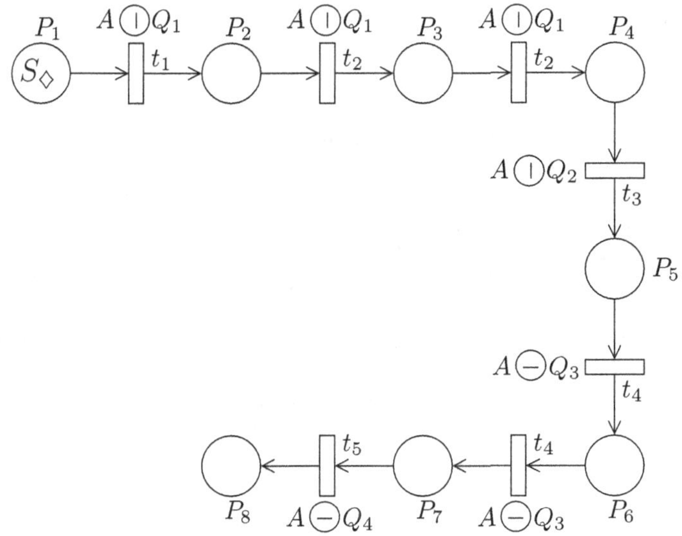

Fig. 9. PATPNS generating PAL-LOC given in Example 5

$1 \leq i \leq k$, $A, Q_i \in \Gamma_p^{**}$, $A', Q_i' \in \Sigma_p^{**}$. I' and O' are input and outplaces of the transitions. P' is the set of places.

Hence we can clearly say that PATPNS is closed under projection.

Theorem 6. *$PAL - REC \subsetneq PATPNS$.*

Proof. $PAL - REC \subseteq PATPNS$ follows from Theorems 4 and 5, since every recognizable partial array language is a projection of a local partial array language.

Now the proper inclusion can be proved easily by giving an example of a partial array language which is not in PAL-REC but in PATPNS.

6 Conclusion

In this paper we have proposed PATPNS and PATPNPS. PATPNS is compared with PAL-LOC, PAL-REC and BPPAG. It is also compared with PATPNPS. The properties of PATPNS and PATPNPS can be studied further by introducing inhibitor arc to increase the generative capacity of PATPNS. This is our future work.

References

1. Berstel, J., Boasson, L.: Partial words and a theorem of Fine and Wilf. Theor. Comput. Sci. **218**(1), 135–141 (1999)

2. Giammarresi, D., Restivo, A.: Two-dimensional languages. In: Rozenberg, G., Salomaa, A. (eds.) Handbook of Formal Languages, pp. 215–267. Springer, Heidelberg (1997). https://doi.org/10.1007/978-3-642-59126-6_4
3. Lalitha, D.: Rectangular array languages generated by a Colored Petri Net. In: IEEE International Conference on Electrical Computer and Communication Technologies, pp. 1–5 (2015)
4. D., L.: Rectangular array languages generated by a Petri net. In: Sethi, I.K. (ed.) Computational Vision and Robotics. AISC, vol. 332, pp. 17–27. Springer, New Delhi (2015). https://doi.org/10.1007/978-81-322-2196-8_3
5. Lalitha, D., Rangarajan, K., Thomas, D.G.: Rectangular arrays and Petri nets. In: Barneva, R.P., Brimkov, V.E., Aggarwal, J.K. (eds.) IWCIA 2012. LNCS, vol. 7655, pp. 166–180. Springer, Heidelberg (2012). https://doi.org/10.1007/978-3-642-34732-0_13
6. Mary Metilda, M.I., Lalitha, D.: Kolam generated by color Petri nets. In: Tuba, M., Akashe, S., Joshi, A. (eds.) Information and Communication Technology for Sustainable Development. AISC, vol. 933, pp. 675–681. Springer, Singapore (2020). https://doi.org/10.1007/978-981-13-7166-0_68
7. Mary Metilda, M.I., Lalitha, D.: Petri nets for pasting tiles. In: Solanki, V.K., Hoang, M.K., Lu, Z.J., Pattnaik, P.K. (eds.) Intelligent Computing in Engineering. AISC, vol. 1125, pp. 701–708. Springer, Singapore (2020). https://doi.org/10.1007/978-981-15-2780-7_76
8. Nivat, M., Saoudi, A., Subramanian, K.G., Siromoney, R., Dare, V.R.: Puzzle grammar and context-free array grammars. Int. J. Pattern Recogn. Artif. Intell. **05**(05), 663–676 (1991)
9. Paun, Gh.: Computing with membranes. J. Comput. Syst. Sci. **61**(1), 108–143 (2000)
10. Peterson, J.L.: Petri Net Theory and Modeling of Systems. Prentice Hall Inc., Englewood Cliffs (1981)
11. Sasikala, K., Kalyani, T., Thomas, D.G.: Partial array grammars and partial array-rewriting *P* systems. Math. Eng. Sci. Aerosp. **11**(1), 227–236 (2020)
12. Subramanian, K.G., Saravanan, R., Geethalakshmi, M., Helen Chandra, P., Margenstern, M.: *P* systems with array object and array rewriting rules. Prog. Nat. Sci. **17**(4), 479–485 (2007)
13. Sweety, F., Sasikala, K., Kalyani, T., Thomas, D.G.: Partial array-rewriting *P* systems and basic puzzle partial array grammar. In: AIP Conference Proceedings, vol. 2277, p. 030003 (2020)
14. Sweety, F., Thomas, D.G., Dare, V.R., Kalyani, T.: Recoginizability of partial array languages. J. Comb. Math. Comb. Comput. **69**, 237–249 (2009)
15. Vijaya Chitra, S., Sasikala, K.: Squares in partial arrays. In: AIP Conference Proceedings, vol. 2112, pp. 20–34 (2019)

Certain State Sequences Defined by P Systems with Reactions

Sastha Sriram[1(✉)], Somnath Bera[2], and K. G. Subramanian[3]

[1] Department of Mathematics, School of Arts, Science and Humanities, SASTRA Deemed University, Tanjore 613 401, Tamil Nadu, India
[2] School of Advanced Sciences-Mathematics, Vellore Institute of Technology, Chennai 600 127, Tamil Nadu, India
[3] School of Mathematics, Computer Science and Engineering, Liverpool Hope University, Hope Park L16 9JD, Liverpool, UK

Abstract. Reaction system was introduced by Ehrenfeucht and Rozenberg [2] as a model of interactions in biochemical reactions while the seminal paper by Gh. Păun [5] introducing the bio-inspired model of P system launched the field of membrane computing. An investigation bridging these two models is done by Păun and Pérez-Jiménez [8]. Here we introduce a variant of transition P system having a finite base set S, called transition P system based on reactions (in short, $(R)TPS$) with the regions of the system having reactions (playing the role of evolution rules) as well as states (in the place of objects) that are subsets (that can be empty) of S. Long terminating state sequences and cycles have been generated in studies of reaction systems. Here we construct $(R)TPS$ generating such sequences with some improvements in the lengths of the sequences.

Keywords: Reaction system · State sequences · Membrane computing · P system

1 Introduction

Reaction system was introduced by Ehrenfeucht and Rozenberg [2] as an abstract model formalizing the interactions among biochemical reactions in living cells with the mathematical formulation taking into account the features of facilitation and inhibition in a reaction. In fact the model has turned out as a general framework for various investigations in different research themes (see, for example, [1–3,9–11] and references therein). On the other hand, in the area of membrane computing, a bio-inspired model, known by the generic name of P system, was proposed by Păun [5]. The basic version of a P system and its several variants have provided a rich platform for handling different kinds of problems related to computing (see, for example, [6,7,12]).

K. G. Subramanian — Honorary Visiting Professor.

R. Freund et al. (Eds.): CMC 2020, LNCS 12687, pp. 153–160, 2021.
https://doi.org/10.1007/978-3-030-77102-7_9

In [8], an investigation bridging the two models of P systems and reaction systems, is done. Motivated by the study in [8], a variant of the basic model of transition P system having a finite base set as in a reaction system, is introduced here. The regions of the proposed variant of P system, referred to as P system based on reactions ($(R)TPS$, in short), can have reactions (playing the role of evolution rules) and also states (as in a reaction system) that are subsets of the base set. Long terminating state sequences and long cycles have been generated [3,9,10] in the context of reaction systems. Here we consider this problem and construct $(R)TPS$ generating such sequences with some improvements in the lengths, corresponding to sequences generated by reaction systems.

2 Basic Definitions

For unexplained notions and notations relating to reaction systems, we refer to [3,9,10] and for notions relating to P systems, we refer to [6]. We first recall reaction system [2].

Definition 1. [2] *Let S be a finite base set.*

(i) A reaction over S is a triple $\rho = (R, I, P)$, where $R \subseteq S$, called reactant set of ρ, $I \subseteq S$, called inhibitor set of ρ, and $P \subseteq S$, called product set of ρ, are nonempty sets such that $R \cap I = \emptyset$;

(ii) Given a reaction $\rho = (R, I, P)$ over S and a subset T of S, the set T is enabled with respect to ρ, if $R \subseteq T$ and $I \cap T = \emptyset$. If T is enabled, then we define the result by $res_\rho(T) = P$; If T is not enabled, then we define $res_\rho(T) = \emptyset$;

(iii) A reaction system $\mathcal{A}_S$ over S (or simply, $\mathcal{A}$ when S is understood) is a finite nonempty set $\{\rho_i/1 \leq i \leq m\}$, of reactions over S; For a reaction system $\mathcal{A} = \{\rho_i/1 \leq i \leq m\}$, we define the result by $res_\mathcal{A}(T) = \bigcup_{i=1}^{m} res_{\rho_i}(T)$; We also write $T \Rightarrow_\mathcal{A} T'$ ($T \Rightarrow T'$ when $\mathcal{A}$ is understood) where $T' = res_\mathcal{A}(T)$ and T, T' are also referred to as states.

Remark 1. Adopting the notation used in [10], we denote the singleton set $\{x\}$ by x and also use the simplified way of writing the set $\{x, y, z\}$ by xyz and the set $\{x, y, z, u, v\}$ by $xyzuv$ and so on, whenever there is no confusion. The cardinality of a set X is denoted by $card(X)$. If, for every reaction $\rho = (R, I, P)$ in a reaction system $\mathcal{A}$, $card(R) \leq r$, $card(I) \leq i$, then $\mathcal{A}$ is called a (r, i) reaction system and the reactions are also called (r, i) reactions. A state sequence (also called, simply, a sequence) $T_0 \Rightarrow T_1 \Rightarrow \cdots \Rightarrow T_m$, $T_i \subseteq S$ in a reaction system with base set S, is either a sequence of length m terminating with $T_m = \emptyset$ or has a cycle of length $m - n$, (i.e) $T_m = T_n$ for some $n, 1 \leq n \leq m-1$ since S is finite.

Example 1. Let the base set $S = \{1, 2, 3, 4\}$.

(i) Let $\mathcal{A}_1$ be a $(1, 1)$ reaction system consisting of $(1, 1)$ reactions $\rho_1 = (1, 3, 4)$, $\rho_2 = (4, 2, 2)$, $\rho_3 = (2, 3, 3)$, $\rho_4 = (3, 2, 123)$. Then $T = 1 \Rightarrow 4$, as the reactant 1 of ρ_1 is a subset of T, the inhibitor 3 of ρ_1 satisfies $3 \cap T = \emptyset$ and the product of ρ_1 is 4. Likewise $4 \Rightarrow 2$, $2 \Rightarrow 3$, $3 \Rightarrow 123$ and $123 \Rightarrow \emptyset$ as 123 is not enabled with respect to any of the reactions. Hence we have a sequence

$$1 \Rightarrow 4 \Rightarrow 2 \Rightarrow 3 \Rightarrow 123 \Rightarrow \emptyset$$

which is terminating and is of length 5. Note that we have used here the simplified notation mentioned in Remark 1. For example, the reaction $\rho_1 = (1, 3, 4)$ is in fact $\rho_1 = (\{1\}, \{3\}, \{4\})$ and likewise $T = 1$ stands for $T = \{1\}$.

(ii) Let $\mathcal{A}_2$ be a $(1, 1)$ reaction system consisting of $(1, 1)$ reactions $\rho_1 = (1, 4, 4)$, $\rho_2 = (4, 3, 2)$, $\rho_3 = (2, 1, 13)$, $\rho_4 = (3, 2, 4)$. Then we have the sequence $T_0 = 1 \Rightarrow T_1 = 4 \Rightarrow T_2 = 2 \Rightarrow T_3 = 13 \Rightarrow T_1 = 4$ having a cycle $T_1 = 4 \Rightarrow T_2 = 2 \Rightarrow T_3 = 13 \Rightarrow T_1 = 4$ of length 3.

3 Transition P System Based on Reactions

The basic model of a transition P system was introduced by Păun [5]. Informally described, a transition P system in its basic version has membranes, hierarchically arranged one within another. There is an outermost membrane, called the skin membrane which contains all other membranes. The membrane having no other membrane inside it, is called an elementary membrane. The regions of the membranes can have objects and evolution rules. The minimal activity in a P system involves processing the objects, if any, in all regions of the system, at the same time. This is done by a nondeterministic and maximally parallel manner of application of the rules to the objects. This allows the objects to evolve and the evolved objects can continue to be in the same region or move to an immediate neighbouring region, with the object communication being decided by a target indication. A computation comes to a halt when no object in all the regions can further evolve and the result of a computation is the number of objects in a specified membrane.

Several modifications and variants of the basic model of a P system have been proposed. We now introduce a variant of transition P system [5], called transition P system based on $(1, 1)$ reactions over a finite base set. Subsets of the base set are the "objects" in the regions of the P system. The P system can have in its regions "tables" of $(1, 1)$ reactions which are sets of reactions.

Definition 2. *A transition P system based on reactions ((R)TPS) is* $\Pi = (S, \mu, Q_1, \cdots, Q_m, T_1, \cdots, T_m)$ *where*

(i) S *is a finite nonempty set, called base set;*
(ii) μ *is the membrane structure with a cell-like hierarchical arrangement of* $m, (m \geq 1)$ *membranes labelled* $1, 2, \cdots, m$ *in a one-to-one manner;*
(iii) each $Q_i, 1 \leq i \leq m$, *consists of a finite number of tables, where a table is a finite nonempty set having a finite number of* $(1, 1)$ *reactions of the form* $\rho = (R, I, P)$ *as defined in Definition 1; each table* t *in a region has a target* $tar \in \{here, out, in\}$ *attached to it, indicated by* $t(tar)$, *with the usual meaning;*

(iv) each $T_i, 1 \leq i \leq m$, called a state, is a distinct subset of S initially in region i and can also be empty.

In any region, if a state $T \subseteq S$ is enabled by reactions in a table t and if T' is the result obtained (as in Definition 1), then we say that the state T' is specified by T and we write $T \Rightarrow T'$; Depending on tar attached to t, the result state T' remains in the same region or is sent to the immediate outer region or the immediate inner region according as tar is here or out or in; At a step, the states, if any, in all the regions define corresponding states. If $T_0 \Rightarrow T_1 \Rightarrow T_2 \Rightarrow \cdots \Rightarrow T_n$ is a state sequence defined in a sequence of steps, then it will either terminate or will have cycle. Note that in this sequence a T_j $(1 \leq j \leq n)$ can be in the same region in which T_0 was there or might have moved to some other region.

We give an example.

Example 2. Consider the $(R)TPS$

$$\Pi_1 = (\{1,2,3,4\}, [_1[_2[_3\ \ ]_3]_2]_1, Q_1, Q_2, Q_3, T_1, T_2, T_3)$$

where
$Q_1 = \{t_1(here), t_2(in)\}$, $Q_2 = \{t_3(in)\}$, $Q_3 = \{t_4(out), t_5(here), t_6(here)\}$;
$T_1 = 1$, $T_2 = T_3 = \emptyset$. The tables are given by

$$t_1 = \{(1,3,2),(2,3,3),(3,1,4)\}, t_2 = \{(4,3,14)\},$$
$$t_3 = \{(1,3,12),(2,4,3),(3,4,4),(4,3,1)\},$$
$$t_4 = \{(1,3,2),(2,3,3)\}, t_5 = \{(3,1,1),(4,2,3)\}, t_6 = \{(1,2,12),(3,4,34)\}.$$

Starting with the state 1 in region 1, the state 1 is enabled by reactions in the table t_1 and so we have $1 \Rightarrow 2 \Rightarrow 3 \Rightarrow 4$. The state 4 remains in region 1 itself and is enabled by reaction in table t_2 to give $4 \Rightarrow 14$. The state 14 is sent to region 2 due to target indication *in* for table t_2. In region 2, $14 \Rightarrow 12$ and the state 12 is sent to region 3. In this region 3, $12 \Rightarrow 23$ and the state 23 is sent back to region 2 where $23 \Rightarrow 34$ and the state 34 is sent to the inner region 3. Here $34 \Rightarrow 13 \Rightarrow 1234 \Rightarrow \emptyset$. The state sequence obtained is $1 \Rightarrow 2 \Rightarrow 3 \Rightarrow 4 \Rightarrow 14 \Rightarrow 12 \Rightarrow 23 \Rightarrow 34 \Rightarrow 13 \Rightarrow 1234 \Rightarrow \emptyset$ which is of length 10.

Long terminating sequences have been obtained in $(1,1)$ reaction systems [3]. Improving the result on long terminating sequence established in [3, Lemma 14, Page 174], Salomaa [9, Lemma 5, Page 280] obtained the following result on long terminating sequence.

Theorem 1. [9] *Assume that the base set S of a $(1,1)$ reaction system $\mathcal{A}$ is of cardinality n and that $\mathcal{A}$ has a terminating state sequence of length $m \geq 1$. Then there is another $(1,1)$ reaction system $\mathcal{A}_1$, with the base set of cardinality $n+4$ and with a terminating state sequence of length $3m+5$.*

Given a $(1,1)$ reaction system $\mathcal{A}$ over a base set of cardinality n and a terminating sequence $T_0 \Rightarrow T_1 \Rightarrow \cdots \Rightarrow T_m$ of length m, here we construct a $(R)TPS$ with a base set of cardinality $n+4$ and having five membranes, defining a terminating sequence of length $5m+9$. The construction is based on the proof technique used in [9, Lemma 5, Page 280].

Theorem 2. *If there is a terminating sequence of length m in a $(1,1)$ reaction system $\mathcal{A}$ with a set A of $(1,1)$ reactions over a base set S of cardinality n, then there is a $(R)TPS$ with a base set of cardinality $n+4$ and having five membranes that defines a terminating sequence of length $5m+9$.*

Proof. Let $X_0 \Rightarrow X_1 \Rightarrow \cdots \Rightarrow X_m = \emptyset, m \geq 1, (X_i \subseteq S, 0 \leq i \leq m-1)$ be a terminating sequence in a reaction system $\mathcal{A}$ over a base set S of cardinality n. Consider the $(R)TPS$

$$\Pi_1 = (S \cup \{a,b,c,d\}), [_1[_2[_3[_4[_5 \quad]_5]_4]_3]_2]_1, Q_1, Q_2, Q_3, Q_4, Q_5, X_0 \cup \{a,b\}, \emptyset, \emptyset, \emptyset, \emptyset)$$

where

$$Q_1 = \{t_1(here), t_2(in)\}, Q_2 = \{t_3(here), t_4(in)\},$$
$$Q_3 = \{t_5(here), t_6(in)\}, Q_4 = \{t_7(here), t_8(in)\}, Q_5 = \{t_9(here)\}.$$

The tables are given by

$$t_1 = A \cup \{\rho_{1i} = (i,d,b) \mid \text{for all} \quad i \in S\} \cup \{\rho_3 = (b,c,a)\},$$
$$t_2 = \{\rho_4 = (a,b,X_0 \cup \{b,c\})\},$$
$$t_3 = A \cup \{\rho_{1i} = (i,d,b) \mid \text{for all} \quad i \in S\} \cup \{\rho_5 = (b,a,c)\},$$
$$t_4 = \{\rho_6 = (c,b,X_0 \cup \{d,a\})\},$$
$$t_5 = A \cup \{\rho_{2i} = (i,b,d) \mid \text{for all} \quad i \in S\} \cup \{\rho_7 = (d,c,a)\},$$
$$t_6 = \{\rho_8 = (a,d,X_0 \cup \{d,c\})\},$$
$$t_7 = A \cup \{\rho_{2i} = (i,b,d) \mid \text{for all} \quad i \in S\} \cup \{\rho_9 = (d,a,c)\},$$
$$t_8 = \{\rho_{10} = (c,d,X_0 \cup \{b\})\},$$
$$t_9 = A \cup \{\rho_{1i} = (i,d,b) \mid \text{for all} \quad i \in S\}$$

where A is the set of all reactions in $\mathcal{A}$. We start with the initial state $X_0 \cup \{a,b\}$ in region 1. There are no states in other regions initially. The state $X_0 \cup \{a,b\}$ is enabled by reactions in table t_1 in region 1. Note that due to the reactions in A and ρ_{1i} in table t_1 in region 1, $X_0 \Rightarrow X_1 \cup b$ while due to reaction ρ_3 in table t_1 in region 1, $b \Rightarrow a$ so that we obtain $X_0 \cup \{a,b\} \Rightarrow X_1 \cup \{a,b\}$ which is retained in region 1 itself as the target of t_1 is *here*. Likewise in region 1, $X_1 \cup \{a,b\} \Rightarrow X_2 \cup \{a,b\}$ and so on. Thus

$$X_0 \cup \{a,b\} \Rightarrow X_1 \cup \{a,b\} \Rightarrow X_2 \cup \{a,b\} \Rightarrow \cdots$$
$$\Rightarrow X_{m-1} \cup \{a,b\} \Rightarrow \{a,b\} \Rightarrow \{a\}$$

since $X_m = \emptyset$ and only due to reaction ρ_3, we have $\{a,b\} \Rightarrow \{a\}$. The state {a} in region 1 is now enabled by ρ_4 in table t_2 so that the state sequence is now

$$X_0 \cup \{a,b\} \Rightarrow X_1 \cup \{a,b\} \Rightarrow X_2 \cup \{a,b\} \Rightarrow \cdots$$
$$\Rightarrow X_{m-1} \cup \{a,b\} \Rightarrow \{a,b\} \Rightarrow \{a\} \Rightarrow X_0 \cup \{b,c\}$$

which is of length $m+2$. The state $X_0 \cup \{b,c\}$ is sent to region 2 due to the target in attached with table t_2. In this region the reactions in table t_3 followed finally by t_4 yield

$$X_0 \cup \{b,c\} \Rightarrow X_1 \cup \{b,c\} \Rightarrow X_2 \cup \{b,c\} \Rightarrow \cdots$$

$$\Rightarrow X_{m-1} \cup \{b,c\} \Rightarrow \{b,c\} \Rightarrow \{c\} \Rightarrow X_0 \cup \{a,d\}$$

which is of length $m+2$. The state $X_0 \cup \{a,d\}$ is sent to region 3 due to the target in attached with table t_4. In this region the reactions in table t_5 followed finally by t_6 yield

$$X_0 \cup \{a,d\} \Rightarrow X_1 \cup \{a,d\} \Rightarrow X_2 \cup \{a,d\} \Rightarrow \cdots$$

$$\Rightarrow X_{m-1} \cup \{a,d\} \Rightarrow \{a,d\} \Rightarrow \{a\} \Rightarrow X_0 \cup \{d,c\}$$

which is of length $m+2$. The state $X_0 \cup \{d,c\}$ is sent to region 4 due to the target in attached with table t_6. In this region the reactions in table t_7 followed finally by t_8 yield

$$X_0 \cup \{d,c\} \Rightarrow X_1 \cup \{d,c\} \Rightarrow X_2 \cup \{d,c\} \Rightarrow \cdots$$

$$\Rightarrow X_{m-1} \cup \{d,c\} \Rightarrow \{d,c\} \Rightarrow \{c\} \Rightarrow X_0 \cup \{b\}$$

which is of length $m+2$. The state $X_0 \cup \{b\}$ is sent to region 5 due to the target in attached with table t_8. In this region the reactions in table t_9 yield

$$X_0 \cup \{b\} \Rightarrow X_1 \cup \{b\} \Rightarrow X_2 \cup \{b\} \Rightarrow \cdots$$

$$\Rightarrow X_{m-1} \cup \{b\} \Rightarrow \{b\} \Rightarrow \emptyset.$$

which is of length $m+1$. The total length of the terminating state sequence is $5m+9$.

In [10, Lemma 8, Page 95], a long terminating sequence from a long cycle was constructed which is recalled in the following result.

Theorem 3. [10] *Assume that the* $(1,1)$ *reaction system* A *with the base set* S *generates a cycle of length* $m \geq 4$. *Then there is a* $(1,1)$ *reaction system* A_1 *with the base set* $S \cup \{a,b,c\}$ *generating a terminating sequence of length* $m+4$.

Given a $(1,1)$ reaction system $\mathcal{A}$ over a base set of cardinality n and a state sequence having a cycle of length m, here we construct a $(R)TPS$ with a base set of cardinality $n+2$, defining a terminating sequence of length $m+2$. The construction is based on the proof technique used in [10, Lemma 8, Page 95].

Theorem 4. *If there is a state sequence having a cycle of length* m *in a* $(1,1)$ *reaction system* $\mathcal{A}$ *with a set* A *of reactions over a base set* S *of cardinality* n, *then there is a* $(R)TPS$ *with a base set of cardinality* $n+2$ *that defines a terminating sequence of length* $m+2$.

Proof. Let $Y_1 \Rightarrow Y_2 \Rightarrow \cdots \Rightarrow Y_m \Rightarrow Y_1, m \geq 1$, $(Y_i \subseteq S, 1 \leq i \leq m)$ be a cycle defined by a $(1,1)$ reaction system $\mathcal{A}$ over a base set S of cardinality n. Consider the $(R)TPS$

$$\Pi_2 = (S \cup \{a,b\}, [_1[_2 \cdots [_{m+1} \quad]_{m+1} \cdots]_2]_1, Q_1, Q_2, \cdots, Q_{m+1}, Y_1, \emptyset, \emptyset, \cdots, \emptyset).$$

For $1 \leq i \leq m$, $Q_i = \{t_i(in)\}$ and $Q_{m+1} = \{t_{m+1}(here)\}$. The tables are given by $t_1 = A, t_2 = A \cup \{(j,b,a)\}$, for j ranging over $S - Y_1$, $t_i = t_2$, for $3 \leq i \leq m$, $t_{m+1} = \{(a,b,S)\}$.

Only the region 1 has initial state Y_1 while other regions do not have any initial state. The state Y_1 in region 1 is enabled by reactions in A and $Y_1 \Rightarrow Y_2$ and the resulting state Y_2 is sent to region 2. Here the additional reactions (j,b,a) (j ranging over $S - Y_1$) yield a and so $Y_2 \Rightarrow Y_3 \cup \{a\}$ which is then sent to region 3. The process continues in regions 3 to m so that $Y_3 \cup \{a\} \Rightarrow Y_4 \cup \{a\} \Rightarrow \cdots \Rightarrow Y_m \cup \{a\} \Rightarrow Y_1 \cup \{a\}$, finally sending $Y_1 \cup \{a\}$ to the innermost region $m+1$ where this state is enabled by the only reaction (a,b,S) so that $Y_1 \cup \{a\} \Rightarrow S \Rightarrow \emptyset$. Thus we have

$$Y_1 \Rightarrow Y_2 \Rightarrow Y_3 \cup \{a\} \cdots \Rightarrow Y_m \cup \{a\} \Rightarrow Y_1 \cup \{a\} \Rightarrow S \Rightarrow \emptyset$$

which is a terminating sequence of length $m+2$.

It was shown in [3, Lemma 16, Page 176] that, by adding three elements to the base set of a reaction system, we get a cycle of length $m+2$ from a terminating sequence of length m. Given a $(1,1)$ reaction system $\mathcal{A}$ over a base set of cardinality n and a state sequence having a terminating sequence of length m, here we construct a $(R)TPS$ with a base set of cardinality $n+3$, defining a cycle of length $m+1$.

Theorem 5. [3] *If a* $(1,1)$ *reaction system generates a terminating sequence of length* m*, one can construct another* $(1,1)$ *reaction system, with three more elements in the base set, generating a cycle of length* $m+2$.

Theorem 6. *If there is a terminating state sequence of length* m *in a* $(1,1)$ *reaction system* $\mathcal{A}$ *with a set* A *of reactions over a base set* S *of cardinality* n*, then there is a* $(R)TPS$ *with a base set of cardinality* $n+3$ *that defines a cycle of length* $m+1$.

Proof. Let $Z_0 \Rightarrow Z_1 \Rightarrow \cdots \Rightarrow Z_m = \emptyset, m \geq 1$, $(Z_i \subseteq S, 1 \leq i \leq m)$ be a terminating sequence defined by a $(1,1)$ reaction system $\mathcal{A}$ over a base set S of cardinality n. Consider the $(R)TPS$

$$\Pi_3 = (S \cup \{a,b,c\}, [_1[_2 \quad]_2]_1, Q_1, Q_2, Z_0 \cup \{a,b\}, \emptyset)$$

where $Q_1 = \{t_1(here), t_2(in)\}$ and $Q_2 = \{t_3(here), t_4(out)\}$. The tables are given by $t_1 = A \cup \{(a,c,b)\} \cup \{(j,c,a) \mid j \in S\}$, $t_2 = \{(b,c,Z_0 \cup \{a,b\})\}$, $t_3 = t_1, t_4 = t_2$. Only the region 1 has initial state $Z_0 \cup \{a,b\}$ while region 2 does not have any initial state. The state $Z_0 \cup \{a,b\}$ in region 1 is enabled by reactions in t_1 and $Z_0 \cup \{a,b\} \Rightarrow \cdots \Rightarrow Z_{m-1} \cup \{a,b\} \Rightarrow \{b\}$. The state $\{b\}$ is enabled by the reaction in table t_2 and the resulting state $Z_0 \cup \{a, b\}$ is sent to region 2. Here the reactions of t_3 yield the state $\{b\}$ and this state is enabled by the reaction in t_4 yielding $Z_0 \cup \{a, b\}$ which is sent back to region 1. The process repeats. Thus we have

$$Z_0 \cup \{a,b\} \Rightarrow Z_1 \cup \{a,b\} \Rightarrow \cdots \Rightarrow Z_{m-1} \cup \{a,b\} \Rightarrow \{b\} \Rightarrow Z_0 \cup \{a,b\}$$

which is a cycle of length $m + 1$.

4 Concluding Remarks

A general question that is considered in the area of membrane computing is the following: Is it possible to reduce the number of membranes used in a P system developed for solving a problem? Here again, one can consider this problem in the P systems constructed in the results on long terminating sequences or on long cycles. Graph-based reaction systems have been introduced and investigated in [4]. It will be of interest, as pointed out in [4], to study sequences of graphs defined by graph-based reaction systems, which will be our future work.

Acknowledgement. The authors are grateful to the reviewers for their very useful comments.

References

1. Brijder, R., Ehrenfeucht, A., Main, M., Rozenberg, G.: A tour of reaction systems. Int. J. Found. Comput. Sci. **22**, 1499–1517 (2011)
2. Ehrenfeucht, A., Rozenberg, G.: Reaction systems. Fundam. Inform. **75**, 263–280 (2007)
3. Ehrenfeucht, A., Main, M., Rozenberg, G.: Functions defined by reaction systems. Int. J. Found. Comput. Sci. **22**, 167–178 (2011)
4. Kreowski, H.J., Rozenberg, G.: Graph transformation through graph surfing in reaction systems. J. Log. Algebraic Methods Program. **109**(100481), 1–22 (2019)
5. Păun, Gh: Computing with membranes. J. Comput. Syst. Sci. **61**(1), 108–143 (2000)
6. Păun, G.: Membrane computing. In: Lingas, A., Nilsson, B.J. (eds.) FCT 2003. LNCS, vol. 2751, pp. 284–295. Springer, Heidelberg (2003). https://doi.org/10.1007/978-3-540-45077-1_26
7. Păun, G., Rozenberg, G., Salomaa, A. (eds.): Handbook of Membrane Computing. Oxford University Press, Oxford (2010)
8. Păun, Gh, Pérez-Jiménez, M.J.: Towards bridging two cell-inspired models: P systems and R systems. Theor. Comput. Sci. **429**, 258–264 (2012)
9. Salomaa, A.: On state sequences defined by reaction systems. Lect. Notes Comput. Sci. **7230**, 271–282 (2012)
10. Salomaa, A.: Functions and sequences generated by reaction systems. Theor. Comput. Sci. **466**, 87–96 (2012)
11. Teh, W.C., Womasuthan, N.: On irreducible reaction systems. Malaysian J. Math. Sci. **12**(1), 25–34 (2018)
12. Zhang, G., Pérez-Jiménez, M.J., Gheorghe, M.: Real-life applications with membrane computing. ECC, vol. 25. Springer, Cham (2017). https://doi.org/10.1007/978-3-319-55989-6

On Numerical 2D P Colonies with the Blackboard and the Gray Wolf Algorithm

Daniel Valenta[1], Miroslav Langer[1,2], Lucie Ciencialová[1,2(✉)], and Luděk Cienciala[1,2]

[1] Institute of Computer Science, Silesian University in Opava, Opava, Czech Republic
{daniel.valenta,miroslav.langer,lucie.ciencialova, ludek.cienciala}@fpf.slu.cz

[2] Research Institute of the IT4Innovations Centre of Excellence, Silesian University in Opava, Opava, Czech Republic

Abstract. The 2D P colonies (see [2]) were introduced as a theoretical model of the multi-agent system for observing the behavior of the community of very simple agents living in the shared environment. Each agent is equipped with a set of programs consisting of a small number of simple rules. These programs allow the agent to act and move in the environment. The 2D P colonies showed to be suitable for the simulations of various (not only) multi-agent systems, and natural phenomena, like the flash floods. The gray wolf algorithm (see [9]) is the optimization-based algorithm inspired by social dynamics found in packs of gray wolves and by their ability to create hierarchies, in which every member has a clearly defined role, dynamically. The wolves' primary goal is to find and hunt down prey, which in our case equals finding the optimal solution to the given problem. The gray wolf algorithm displays positive results thanks to the principles of randomness and communication between wolves. In this paper, we follow our previous research on the numerical 2D P colony with the blackboard (see [12,13]). We present the results of the computer simulation of numerical 2D P colonies and we compare these results with original gray wolf algorithm.

Keywords: Pack algorithm · 2D P colony · Optimization · Multi-agent system · Blackboard

1 Introduction

The 2D P colonies (see [3,4,8]), as a version of the P colonies with a two-dimensional environment, were designed for the purpose of simulating the behavior of the community of very simple agents living in the shared two-dimensional environment. Both these models origin from P systems - biologically motivated theoretical model inspired by function and behaviour of living cells (for more details see [8]). 2D P colonies has shown to be suitable for simulating various simple multi-agent systems, and natural phenomena, like the flash floods(see

R. Freund et al. (Eds.): CMC 2020, LNCS 12687, pp. 161–177, 2021.
https://doi.org/10.1007/978-3-030-77102-7_10

[1]). The multi-agent systems suitable for simulating by the 2D P colonies have to be very simple, like colonies of ants, etc. The only communication device the agents of the 2D P colony can use is the environment, where the agent can leave special symbols for the others. This form of communication corresponds to the stigmergy. When the communication between the agents via the environment is not sufficient, the 2D P colonies are not suitable for simulating such a system.

Also P systems were successfully used for solving optimization problems. Recently, T. Y. Nishida designed membrane algorithms (see [10]) for solving NP-complete optimization problems, namely the traveling salesman problem (see [11]). G. Zhang, J. Cheng and M. Gheorghe proposed ACOPS, the combination of the P systems with ant colony optimization for solving the traveling salesman problems (see [15]). In [14] the similarities between distributed evolutionary algorithms and membrane systems for solving continuous optimization problems were studied.

Numerical P systems (see [7]), which have been used (as well as P colonies) as a robotic controller (in [5]), also work on a similar principle as numerical 2D P colonies.

The gray wolf optimization algorithm (GWO) (see [13]) is already well–established meta-heuristic optimization technology. It is inspired by social dynamics found in packs of gray wolves, and by their ability to create hierarchies, in which every member has a clearly defined role, dynamically. The primary goal of the wolves is to find and hunt down prey in their environment. The algorithm is inspired by the hunting process. The environment is represented by a mathematical fitness function characterizing the problem and the prey represents the extreme.

The original model of the 2D P colony cannot successfully simulate the GWO, while the agents are able to communicate only via the environment, they are not able to share their po5sitions, hence, they are not able to form desired hierarchy, and successfully hunt down the prey. Moreover, the environment of the 2D P colony is a multiset of the objects. In [12], we have introduced the numerical 2D P colony with the blackboard. The environment is represented by the discrete values of the fitness function and the model is equipped with the universal communication device, the blackboard.

We have organized the rest of this paper in the following way: In the second section, we present the gray wolf algorithm. The third section is devoted to the numerical 2D P colonies and the issue of simulation of gray wolf algorithm by this type of 2D P colonies. The receivers are also presented in the last part of the third section. The receivers aim to help estimate the location of each agent in the environment. In the fourth section, we present the results of the computer simulation and we compare the results with the original GWO algorithm. We conclude the paper by the recapitulation of achieved results and discussion about them.

2 Gray Wolf Optimization Algorithm

The gray wolf optimization algorithm (GWO) (see [9]) is already well–established meta-heuristic optimization technology. The gray wolves create a social hierarchy in which every member has a clearly defined role. Each wolf can fulfill one of the following roles:

- *Alpha* pair is the dominant pair and the pack follows their lead for example during hunts or while locating a place to sleep.
- *Beta* wolves support and respect the Alpha pair during its decisions.
- *Delta* wolves are subservient to Alpha and Beta wolves, follow their orders, and control Omega wolves. Delta wolves divide into *scouts* – they observe the surrounding area and warn the pack if necessary, *sentinels* – they protect the pack when endangered and *caretakers* – they provide aid to old and sick wolves.
- *Omega* wolves help to filter the aggression of the pack and frustrations by serving as scapegoats.

The primary goal of the wolves is to find and hunt down prey in their environment. The hunting technique of a wolf pack can be divided into 5 steps:

1. *Search for the prey* – wolves are attempting to find the most valuable prey with respect to the effort required to hunt it successfully.
2. *Exploitation of the prey* – wolves are attempting to draw attention to themselves and to separate the prey from its herd.
3. *Encircling prey* – the attempt to push the prey into a situation from which it cannot escape.
4. *The prey is surrounded* – it can no longer escape.
5. *The attack* – wolves attack the weak spots of the prey (belly, legs, snout) until it succumbs to fatigue. Afterwards, they bring it down and crush its windpipe.

The gray wolf optimization algorithm is inspired by this process and smoothly transitions between scouting and hunting phases. The prey represents the optimal solution to the given problem, and the environment is represented by a mathematical fitness function characterizing this problem. The value of the function at the current position of the particular wolf represents the highest-quality prey. The wolf with the best value is ranked as Alpha, the second one as Beta, third as Delta, and all the others are Omegas.

In the scouting phase, the pack extensively scouts its environment through many random movements so that the algorithm does not get stuck in a local extreme, while in the hunting phase, the influence of random movements is slowly reduced and pack members draw progressively closer to the discovered extreme. To maintain the divergence between those phases, each wolf is assigned vectors $\vec{A}$ and $\vec{C}$.

$$\vec{A} \text{ is a vector with components } rand\ (-1, 1) * a,$$

where $rand(-1, 1)$, generates a random number between -1 and 1 and where

$$a = 2 - \left(\frac{2i}{i_{max}}\right),$$

while i is the current iteration of the algorithm, and i_{max} is the maximum number of iterations. The vector $\vec{A}$ is random value between -2 and 2. The impact of the vector $\vec{A}$ can be seen in the Fig. 1. With growing iterations, it is more likely that the value of the vector will be between -1 and 1, and this increases the probability of the wolf to be hunting.

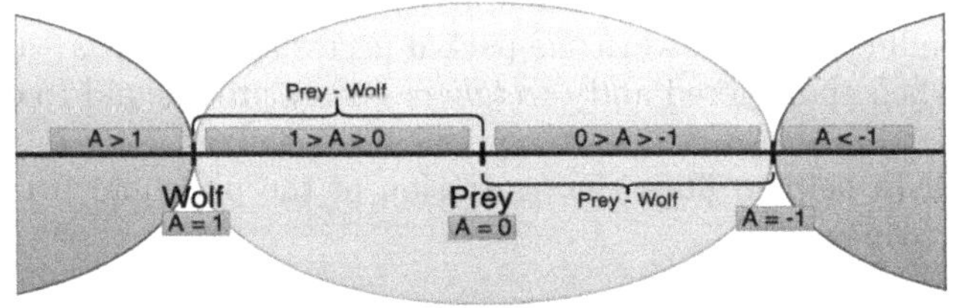

Fig. 1. The vector $\vec{A}$ and its impact in 1D

Another component supporting the scouting phase is the vector

$$\vec{C} = rand(0, 2),$$

where $rand(0, 2)$, generates a random number between 0 and 2. Unlike the vector $\vec{A}$, the vector $\vec{C}$ is not influenced by the iterations. This vector helps the wolves behave more naturally. Analogously, in nature, wolves encounter various obstacles that prevent them from approaching prey comfortably.

The vectors $\vec{A}$ and $\vec{C}$ encourage wolves to prefer scouting or hunting, and so to avoid local optima regardless of the current iteration of the algorithm.

The position of the wolves in the environment are updated on the base of the estimated location of the prey using Alpha, Beta, and Delta wolves as guides.

Let $\vec{X_j}(i)$ be a position vector of wolf j in i-th iteration. The position vector of wolf j is updated as follows:

$$\vec{X_j}\,(i+1) = \frac{\vec{X_1} + \vec{X_2} + \vec{X_3}}{3},$$

where i is the current iteration of the algorithms, and $\vec{X_1}$, $\vec{X_2}$, $\vec{X_3}$ are new potential position vectors of Alpha, Beta, and Delta wolves obtained from following formulas:

$$\vec{X_1} = \vec{X_\alpha}\,(i) - \vec{A_1} * \vec{D_\alpha}$$
$$\vec{X_2} = \vec{X_\beta}\,(i) - \vec{A_2} * \vec{D_\beta}$$
$$\vec{X_3} = \vec{X_\delta}\,(i) - \vec{A_3} * \vec{D_\delta}$$

where $\vec{X_\alpha}(i)$, $\vec{X_\beta}(i)$, $\vec{X_\delta}(i)$ are the position vectors of Alpha, Beta, and Delta wolves, they are representing the positions in the environment that are closest to the optimum in i-th iteration. The vectors $\vec{A_1}, \vec{A_2}, \vec{A_3}$ are calculated in the same way as vector $\vec{A}$. The vectors $\vec{D_\alpha}, \vec{D_\beta}, \vec{D_\delta}$ are defining the distance of the wolf j position from the prey as follows:

$$\begin{aligned}
\vec{D_\alpha} &= \left|\vec{C_1} * \vec{X_\alpha}(i) - \vec{X_j}(i)\right| \\
\vec{D_\beta} &= \left|\vec{C_2} * \vec{X_\beta}(i) - \vec{X_j}(i)\right| \\
\vec{D_\delta} &= \left|\vec{C_3} * \vec{X_\delta}(i) - \vec{X_j}(i)\right|
\end{aligned}$$

where $|\vec{X}|$ is the vector whose components are the absolute values of the components of $\vec{X}$.

The vectors $\vec{C_1}, \vec{C_2}, \vec{C_3}$ are computed in the same way as vector $\vec{C}$, and they influence the weight of the estimated position of the prey $\vec{X_\alpha}, \vec{X_\beta}, \vec{X_\delta}$, increasing or decreasing it.

This principle ensures, that the wolves have the tendency to approach the prey from the different directions and encircle it (see Fig. 2).

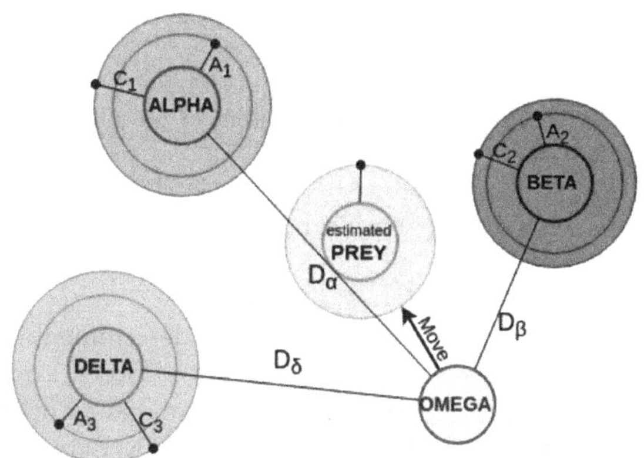

Fig. 2. Update of the positions of the Omega wolves [9].

2.1 Algorithm Pseudo-code

In this subsection we describe the algorithm in pseudo-code.

The inputs of the algorithm are dimensions of the environment of the problem, the boundaries of the environment of the problem, fitness function characterizing the problem, the size of the pack (number of wolves/agents), number of iterations of the algorithm, termination criteria and criteria of the fitness function.

The pseudo-code of the algorithms:

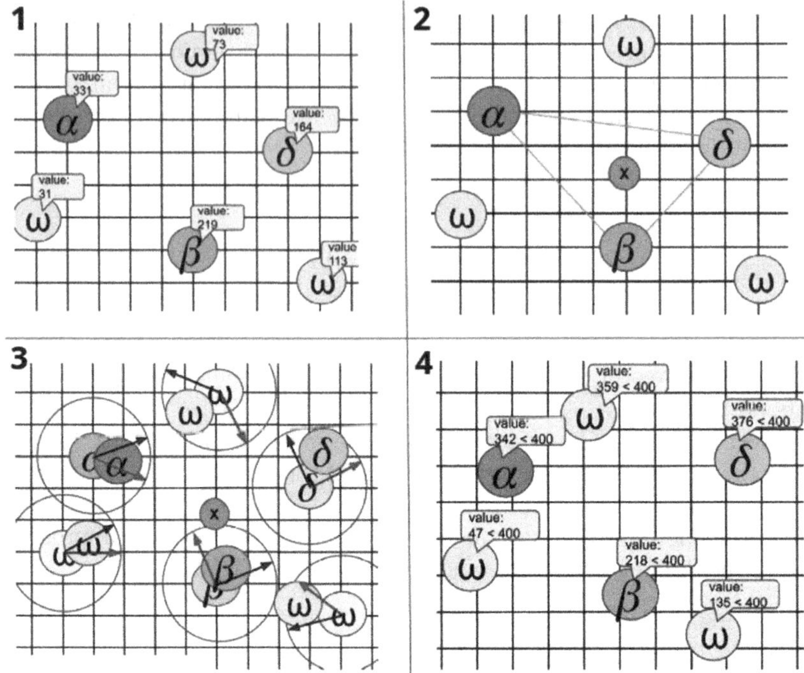

Fig. 3. The visualization of the steps of the algorithm

1. In the first step, agents (wolves) are randomly spread out across the environment.
2. In each iteration i:
   (a) calculate the fitness value of each agent and determine its social hierarchy – Fig. 3. part 1. The agent with the best value (closest to the optimum) is Alpha, second best is Beta, third best is Delta, and all others are Omegas.
   (b) calculate the best solution found so far by Alpha, Beta and Delta ($\vec{X_\alpha}(i)$, $\vec{X_\beta}(i)$, $\vec{X_\delta}(i)$) and average it – Fig. 3. part 2,
   (c) update positions of all the wolves $X_j(i+1)$, while random vectors $\vec{A}$ and $\vec{C}$ are updated for each one – Fig. 3. part 3,
   (d) check the termination criterion – Fig. 3. part 4. Iterations terminate when fitness function value reaches a preset value.

3 The Simulation of the Gray Wolf Optimization Algorithm Using 2D P Colonies

To successfully simulate GWO by the 2D P colony, we need to solve 3 basic problems stated in the Table 1. For more details see [12].

Table 1. Differences between Gray wolf algorithm and 2D P colony

Difference / System	Gray wolf algorithm	P colony	Solution
Environmental problem	The environment is represented by a mathematical fitness function	The environment is represented by a multiset of symbols	Include discretized values of the fitness functions into the environment
Communication problem	The agents have the knowledge of their global position in the environment	They are communities of simple reactive agents independently living and acting in a joint shared environment	The blackboard was introduced
Randomness problem	Random vectors $\vec{A}$ and $\vec{C}$ influence the movement of wolves in the environment	Each rule is deterministic, the only way to implement randomness is to randomize the choosing rule for identical configurations	Non-deterministic choice between several applicable rules

3.1 Model of Numerical 2D P Colony with the Blackboard

Let us recall the definition of the numerical 2D P colony with the blackboard.

Definition 1. *A numerical 2D P colony with blackboard is a construct* $\Pi = (V, e, Env, A_1, \ldots, A_k, BB, f), k \geq 1$, *where*

- *V is the alphabet of the colony. The elements of the alphabet are called objects.* $\square$ *are special objects, that can contain an arbitrary number.*
- *$e \in V$ is the basic environmental object of the numerical 2D P colony,*
- *Env is a triplet $(m \times n, w_E, f_E)$, where $m \times n, m, n \in N$ is the size of the environment. w_E is the initial contents of the environment, it is a matrix of size $m \times n$ of multisets of objects over $V - \{e\} \cup \{f_E(x)\}$. f_E is an environmental function.*
- *A_i, $1 \leq i \leq k$, are the agents. Each agent is a construct $A_i = (o_i, P_i, [o, p])$, $0 \leq o \leq m$, $0 \leq p \leq n$, where*
  - *o_i is a multiset over V, it determines the initial state (contents) of the agent, $|o_i| = 2$,*
  - *$P_i = \{p_{i,1}, \ldots, p_{i,l_i}\}, l \geq 1, 1 \leq i \leq k$ is a finite set of programs, where each program contains exactly 2 rules. Each rule is in the following form:*

   * $a \to b$, *the evolution rule*, $a, b \in V$,
   * $c \leftrightarrow d$, *the communication rule*, $c, d \in V$,
   * $[a_{q,r}] \to s, a_{q,r} \in V, 0 \leq q, r \leq 2, s \in \{\Leftarrow, \Rightarrow, \Uparrow, \Downarrow\}$, *the motion rule*,
   * $a \rightarrowtail \boxed{x}$, $x \in \mathbb{R}, a, \boxed{x} \in V$, *is the blackboard communication rule.*

  If the program contains evolution or communication rule r_1, r_2 *that each works with objects with numbers, it can be extended by a condition:* $\langle x > y \ : \ r_1, r_2 \rangle$, $\langle x \geq y \ : \ r_1, r_2 \rangle$,
- $[o, p]$, $1 \leq o \leq m$, $1 \leq p \leq n$, *is an initial position of agent* A_i *in the 2D environment,*

- BB *is the blackboard.*
- $f \in V$ *is the final object of the colony.*

A configuration of the numerical 2D P colony with the blackboard is given by the state of the environment - matrix of type $m \times n$ with pairs - multiset of objects over $V - \{e\}$, and a number - as its elements, and by the states of all agents - pairs of objects from the alphabet V, and the coordinates of the agents. An initial configuration is given by the definition of the numerical 2D P colony with the blackboard.

A computational step consists of three steps. In the first step, the set of the applicable programs is determined according to the current configuration of the numerical 2D P colony with the blackboard. In the second step, one program from the set is chosen for each agent, in such a way that there is no collision between the communication rules belonging to different programs. In the third step, chosen programs are executed, the values of the environment and on the blackboard are updated. If more agents execute programs to update the same part of blackboard, only one update information is non-deterministically chosen. The agent has no information if his update attempt was successful or not.

A change of the configuration is triggered by the execution of programs, and updating values by functions. It involves changing the state of the environment, contents and placement of the agents.

A computation is non-deterministic and maximally parallel. The computation ends by halting when there is no agent that has an applicable program.

The result of the computation is the number of copies of the final object placed in the environment at the end of the computation.

3.2 Numerical 2D P Colony with the Blackboard for GWO

In this section, we present the numerical 2D P colony with the blackboard that models GWO.

$P_{gw} = (V, e, Env, A_1, A_2, \ldots, A_k, BB, f)$, $k \geq 0$, where:

- $V = \left\{\Box, \Box', \Box'', \Box''', \Box^{iv}, \Box^{v}, \Box^{vi}, e, f, a', b, c, d, h, h', h'', m_{OK}\right\} \cup$
  $\cup \{m_{KO}, m'_{KO}, m''_{KO}, n, A, B, D, \} \cup \{l, l', l'', l''', l^{iv} \mid Y \in \{\Leftarrow, \Rightarrow, \Uparrow, \Downarrow\}\} \cup$
  $\cup \{k_{z_1 z_2 z_3 z_4} \mid z_i \in \{\Leftarrow, \Rightarrow, \Uparrow, \Downarrow\} \wedge k, i \in \{1, 2, 3, 4\}\}$,
- $e \in V$ is the basic environmental object,

- Env is a triplet $(i \times j, w_e, f_E(x, y))$, where $i, j \in \mathbb{N}$, $w_E = |a_{r,s}|, a_{r,s} = \varepsilon, 1 \geq r \geq i, 1 \geq s \geq j$,
- $A_1, A_2, \ldots, A_k$ are the agents, $A_i = (O_i, P_i, [r_x, r_y])$, where:
  - $|o_i| = 2$,
  - $P_1 = P_2 = \cdots = P_k$, P_i rules are defined below,
  - $[r, s]$ are the initial coordinates,
- BB is the blackboard. The blackboard is described in the separate subsection.
- f is the final object, $f \in V$.

The initial configuration of the agent is ee, and its position is $[r, s]$.
The set of programs P_i is as follows:

1. $\langle e \rightarrowtail \boxed{x}'; \ e \to Get(BB[alpha], \boxed{})\rangle$, where $x \in \mathbb{R}$ is number placed in the environmental cell $[r, s]$, $alpha$ is the address of current value y of the alpha wolf on the blackboard. This program reads the number from the environmental cell, and the value of current alpha wolf on blackboard. If more agents decide to be Alpha (or Beta or Delta) in one computational step, The value of only one of them is placed into Blackboard. The computational phase of Alpha, Beta and Delta agents (from configuration ee to configuration ee) is shorter than this phase for Omega agents and in following phase they check the value of three leading wolves again.
2. The programs in the following subset of programs compares two numbers stored inside the agent, and serves to set the role of the agent:
   (a) $\left\langle x > y \ : \ \boxed{x} \to \boxed{x}, \boxed{y}' \to A\right\rangle$ – I am new Alpha,
   (b) $\langle x \geq y : \boxed{x}' \to b, \boxed{x} \to \boxed{x}\rangle$
   (c) $\langle \boxed{x} \to \boxed{x}, b \leftarrow Get(BB[beta], \boxed{}'')\rangle$,
   (d) $\left\langle x > y \ : \ \boxed{x} \to \boxed{x}, \boxed{y}'' \to B\right\rangle$ – I am new Beta,
   (e) $\langle x \geq y : \boxed{x}'' \to c, \boxed{x} \to \boxed{x}\rangle$
   (f) $\langle \boxed{x} \to \boxed{x}, c \leftarrow Get(BB[delta], \boxed{}''')\rangle$,
   (g) $\left\langle x > y \ : \ \boxed{x} \to \boxed{x}, \boxed{y}''' \to D\right\rangle$ – I am new Delta,
   (h) $\langle x \geq y : \boxed{x}''' \to d, \boxed{x} \to \boxed{x}^{iv}\rangle$ – I am Omega,
3. If there is A, B or C inside the agent, then the agent updates the blackboard using the following programs:
   (a) $\langle Update(\boxed{x}, BB[alpha]), A \to a'\rangle$,
   (b) $\langle Update(\boxed{x}, BB[beta]), B \to a'\rangle$,
   (c) $\langle Update(\boxed{x}, BB[delta]), D \to a'\rangle$,
   (d) $\langle \boxed{x} \to e, a' \to e\rangle$,
4. If the agent is the omega wolf, it reads its distance from the prey (the distance is computed by the function and placed on the blackboard), and it moves in a random direction. The direction is generated in such a way, that the agent creates the object 1 with a low index formed from four directions in random order. 1 means that the agent will move in the first direction.
   (a) $\langle d \to 1_w, \boxed{x}^{iv} \to Get(BB[my\ dist.\ from\ prey], \boxed{}^{v})\rangle$, $w = z_1z_2z_3z_4$, $z_i \in \{\Rightarrow, \Leftarrow, \Uparrow, \Downarrow\}$, $z_1 \neq z_2 \neq z_3 \neq z_4$

(b) $\langle 1_w \leftrightarrow e, \boxed{x}^v \rightarrow \boxed{x}^v \rangle$,

(c) $\left\langle \begin{bmatrix} e & e & e \\ e & X_{z_1z_2z_3z_4} & e \\ e & e & e \end{bmatrix} \rightarrow z_X, e \rightarrow z_X \right\rangle$, $X \in \{1,2,3,4\}$, $z_X \in \{\Rightarrow, \Leftarrow, \Uparrow, \Downarrow\}$

Then the agent puts the object with movement sequence into environmental cell and reads its distance from the prey.

(a) $\langle z_X \rightarrow z'_X, \boxed{x}^v \leftrightarrow e \rangle$,

(b) $\langle z'_X \rightarrow z''_X, e \rightarrow h \rangle$,

(c) $\left\langle z''_X \leftrightarrow \boxed{x}^v, h \rightarrow Get(BB[dist.\ from\ prey], \boxed{}^{vi}) \right\rangle$,

5. If the new value is smaller than the value from the previous location, the agent consumes the object corresponding to the direction, and the object with movement sequence and rewrites its content to ee.

   (a) $\left\langle x \geq y : \boxed{x}^v \rightarrow m_{OK}, \boxed{y}^{vi} \rightarrow h' \right\rangle$,

   (b) $\left\langle x > y : \boxed{x}^{vi} \rightarrow m_{KO}, \boxed{y}^v \rightarrow h' \right\rangle$,

   (c) $\langle m_{OK} \rightarrow m_{OK}, h' \leftrightarrow z''_X \rangle$,

   (d) $\left\langle \begin{bmatrix} e\,e\,e \\ e\,e\,e \\ e\,e\,e \end{bmatrix} \rightarrow \overline{z}_X, z''_X \rightarrow z'''_X \right\rangle$, $\overline{z}_X$ is the opposite movement to z_X,

   (e) $\langle m_{OK} \leftrightarrow X_w, z'''_X \rightarrow z^{iv}_X \rangle$,

   (f) $\left\langle \begin{bmatrix} e\,e\,e \\ e\,e\,e \\ e\,e\,e \end{bmatrix} \rightarrow z_X, z^{iv}_X \rightarrow h'' \right\rangle$

   (g) $\langle X_w \rightarrow e, h'' \rightarrow e \rangle$,

6. If the new distance is greater than the old one, then the agent consumes the object corresponding to the direction, and moves to the original location, and rewrites the object 1 with the movement sequence into the object 2 with the same movement sequence, and the agent continues with the investigation. If the agent moves back and there is object 4 with movement sequence, it did not find a smaller distance to the prey and it stops working.

   (a) $\langle m_{KO} \rightarrow m_{KO}, h' \leftrightarrow z''_X \rangle$,

   (b) $\langle m_{KO} \rightarrow m'_{KO}, z'''_X \rightarrow e \rangle$,

   (c) $\langle m'_{KO} \rightarrow m'_{KO}, e \leftrightarrow X_w \rangle$,

   (d) $\langle m'_{KO} \rightarrow m''_{KO}, X_w \rightarrow (X+1)_w \rangle$, $X \in \{1,2,3\}$

   (e) $\langle m''_{KO} \rightarrow Get(BB[my\ dist.\ from\ prey], \boxed{}^v), X_w \leftrightarrow e \rangle$, $X \in \{2,3,4\}$,

   (f) $\langle m'_{KO} \rightarrow f, e \leftrightarrow 4_w \rangle$,

The computation is possibly non-halting because three agents (alpha, beta, and gamma) can always find an applicable program. The positions of these three agents determine the position of the prey.

Blackboard. The blackboard for *GWO* is a structure defined as follows:

$$BB = (\vec{fnc}, [\vec{v1}, \vec{v2}]),\ \text{where:}$$

- dimension of both vectors $\vec{v1}, \vec{v2}$ is $j = \max\{7, k\}$, $k \geq 1$ is the number of agents. In this case, it is a matrix of type $i \times j$, $i, j \in \mathbb{N}$ represented by a vector of these vectors.
- $\vec{v1}$ is a vector with elements that can be named *AlphaValue, BetaValue, DeltaValue, AlphaPosition, BetaPosition, DeltaPosition, preyPosition.* If $j > 7$, then the elements with an index greater than 7 are without a name, they are addressed by its position. The first three elements are serviced by the agents, so the function of the blackboard for the first row is only to copy their values if they are not updated by agents in the current step of the computation.
  - initial content of $\vec{v1}$ is 0 in each element.
- $\vec{v2}$ is a vector with elements named A_0*'s DistanceFromPrey*, A_1*'s DistanceFromPrey*, ..., A_k*'s DistanceFromPrey*, k is number of agents (wolves), the elements without a name can be addressed only by its position in the blackboard matrix.
  - initial content of $\vec{v2}$ is 0 in each element.
- fcn is a vector of functions $(fnc1(i), fnc2(i))$ for manipulating the vectors $\vec{v1}$ and $\vec{v2}$, where $fnc1(i)$ updates $i - th$ element of vector $\vec{v1}$ and $fnc2(i)$ updates $i - th$ element of vector $\vec{v2}$, $0 \leq i \leq j$, j is a dimension of vectors $\vec{v1}$ and $\vec{v2}$:

  i $$fnc1(i) = \begin{cases} identity & i = \{0, 1, 2\}, \\ B_{Position} & i = \{3, 4, 5\}, \\ \frac{fnc1(3)+fnc1(4)+fnc1(4)}{3} & i = 6. \end{cases}$$

  ii $fnc2(i) = |fnc1(preyPosition) - B_{Position}|$, for $i =$ index of agent A_i.

  Auxiliary function $B_{Position}$ is described in the section Receivers.

3.3 Receivers - From Theoretical Model into Real Life

A communication between agents and blackboard is realized by receivers. We equip our model with two receivers that are listening signals coming from the agents. The abilities of receivers are crucial for functioning of the functions of the blackboard, because they are providers of values of $B_{position}$ function. We can assume, that receivers can "see" the position of each agent but for wide areas, it

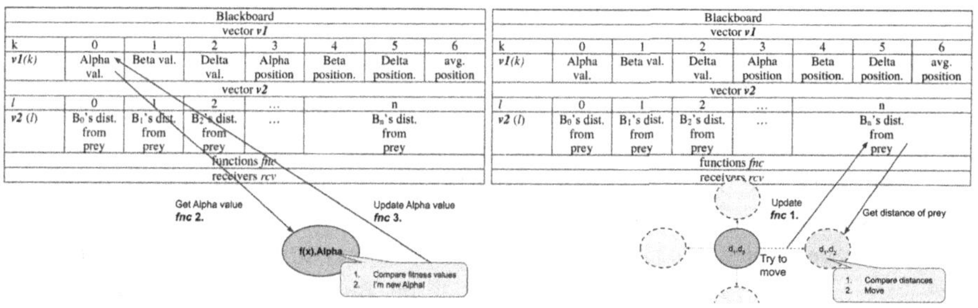

Fig. 4. The use of the blackboard

is not very realistic. For our model, we choose another approach. We introduce time into our model - it takes some time to signal from agents to reach receivers.

- *rcv* - the blackboard has two receivers. Both receivers are located in the environment. Their initial positions are on opposite sides of the boundary points of the environment *Env* (positions $[0, 0]$ and $[m-1, n-1]$, where $m \times n, m, n \in \mathbb{N}$, is the size of the environment).
  The positions of the receivers are updated in each derivation step and receivers circle around the environment as follows:
  - $x < 1, y < n : x = x, y = y + 1,$
  - $x < m, y > n - 1 : x = x + 1, y = y,$
  - $x > m - 1, y > 0 : x = x, y = y - 1,$
  - $x > 0, y < n + 1 : x = x - 1, y = y.$

  The primary task of the *rcv* is to collect data (messages) from agents.
  - The messages have a given structure:

    $$msg : (contents,\ index\ of\ the\ agent,\ timestamp)$$

    Contents is a request of the agent - function *Get*, *Update* or it is an empty string, *timestamp* corresponds to the time, when request was sent.

  Receivers are listening to agents' signal. If both receivers receive the same message from the agent, the received message is being processed in the following sequence: computation of auxiliary function $B_{Position}$, execution of *contents* part of message.
  - $B_{Position} = x \in R_1 \cap R_2$, where R_1, R_2 are circles with center at the positions of the receivers and radius $r = now - sent$, where *now* is the time of receiving the message, and *sent* equals to the *timestamp*, x is chosen randomly in the intersection area. The intersection shapes are changing in the time due to the movements of the receivers.

At this point, it is important to focus on the use of the blackboard by the agents.

If the agent concludes that it is the Alpha, it rewrites the field $\vec{v1}[0]$ using the communication program 3 (Fig. 4, left side). In the same way, Beta and Delta wolves can rewrite field $\vec{v1}[1]$ using the same function. On the right side of the Fig. 4, the agent concludes that it is the Omega, and it will try to move with the assistance of the blackboard, using communication program 1.

4 Computer Simulation

In this section, we present the results of the computer simulation and we compare the results with the original GWO algorithm. We built the simulator based on the analysis from the previous section.

The first version of the simulator was not acting as desired, but the complications were expected. In the first derivation steps, most of the wolves tried to be the Alpha wolf, and they tried to write to the alpha position on the blackboard.

This behavior of the simulator led to the situation when the Omega wolves were without the lead and they reached the final configuration soon, and the simulation stopped unsuccessfully very often, giving no result.

Considering these results, we decided to make a slight change in the behavior of the wolves in the first steps of the derivation. Once the wolf tries to write to the alpha position unsuccessfully, it is allowed to try to write to the beta and delta position respectively. This modification is still in the scope of the P colonies, and we were not forced to introduce some outer mechanism setting the hierarchy of the pack. The wolves were able to set the hierarchy themselves, and the simulation proceeded as expected.

The following figures show the simulator, the settings, the process of the simulation, and the results.

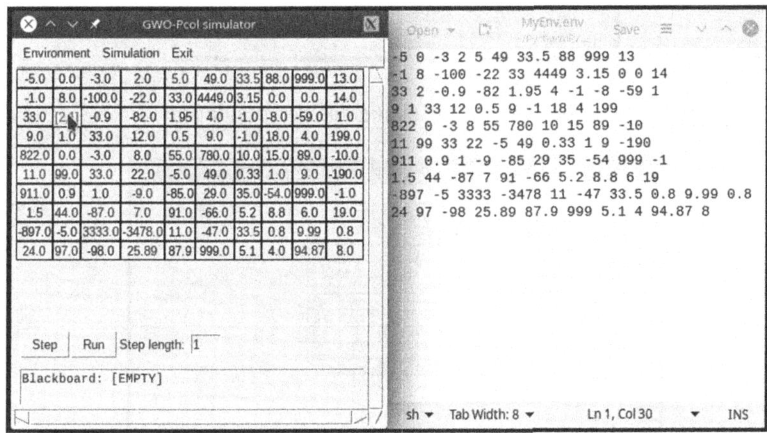

Fig. 5. The instance of the environment

In the Fig. 5, there can be seen an instance of the environment. Source environment file is displayed on the right. If we move the mouse cursor to some field, we obtain the coordinates of it.

In the Fig. 7, the agents are initialized within the environment. The vectors of the blackboard have all their values initialized to zero. The agents are represented by the blue text color. By moving the mouse cursor to the position of the agent, we obtain detailed information about it in the form: $A_i = (\{obj_1, obj_2\}, [posX, posY])$, where i is the index of the agent, obj_1 and obj_2 are the objects inside this agent, and $posX$ and $posY$ are the coordinates of its position.

The *Step* button runs a predefined number of the steps of the simulation. In the first step, each agent runs one rule of its program. The iteration is a set of these steps. New iteration starts when each agent performed all the steps that are needed before the synchronization. The Output console displays the following information: Current iteration, the action of the agent in the form:

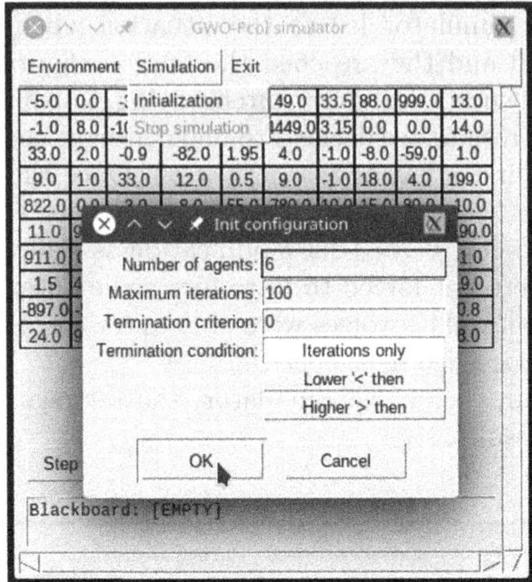

Fig. 6. The initial settings of the simulation

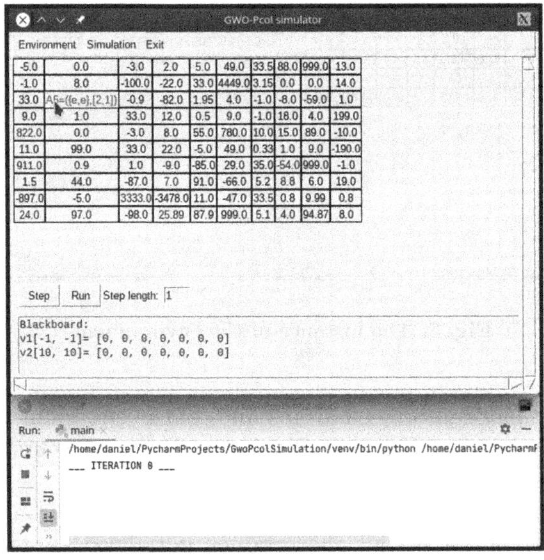

Fig. 7. The agents in the environment (Color figure online)

$(index, (posX, posY), value, action)$ – for example: $(1, (1, 5), 4449, \text{'}A\text{'})$ = agent 1 at the position (1,5) updates Alpha on the blackboard with value 4449, $(4, (2, 6), \text{'}L\text{'}, \text{'}m\text{'})$ = agent 4 at the position (2,6) is moving to position Left (Fig. 8).

The simulation terminates when the termination criterion, the iteration reaches the maximal value, or when no more delta agent can move. In the Fig. 9,

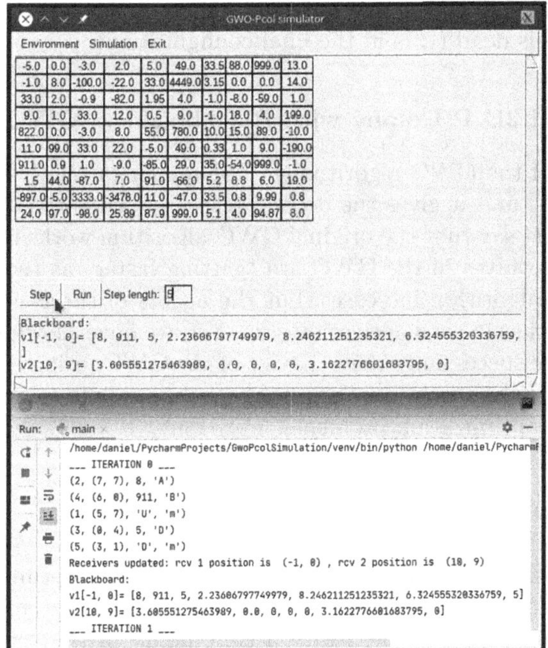

Fig. 8. The simulation

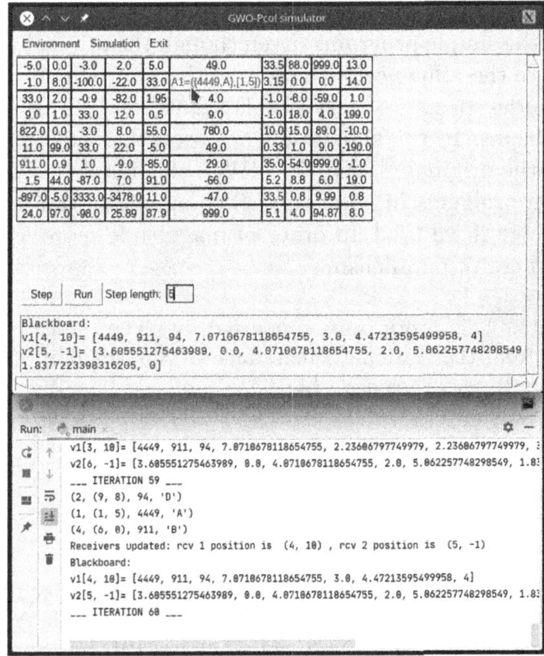

Fig. 9. The last iteration

there can be seen that agent one has found the best solution – number 4449, and that agent 5 is dead (it is in the final configuration).

4.1 Numerical 2D P Colony with Blackboard vs GWO

The simulation of the GWO algorithm by the numerical 2D P colony with the blackboard works, and it gives the desired results, it finds the optimal solution However, one must say that the original GWO algorithm works faster. The main issue is, that the wolves in the GWO are "getting faster" as the number of the iterations of the algorithm increases, but the agents of the P colony can move only one step further in each derivation step. Yet, we have shown that our model is able to simulate successfully the behavior of the GWO, using only simple rules formed into the simple programs, and that even so simple theoretical model is able to solve the optimization problems.

5 Conclusion

The gray wolf optimization algorithm is a well-established optimization method, and the 2D P colonies are already well-established theoretical models. Both are inspired by nature and both deal with a community of the agents acting in a shared environment.

We have shown that by solving the communication problem of the 2D P colony by introducing the blackboard into the model, we are able to simulate successfully the behavior of the pack of wolves of the GWO, using only simple rules formed into the simple programs. Even though the original GWO algorithm works faster, due to the influence of the step of the iteration on the movement of the wolves, our model consists of very simple agents, and still is able to solve the optimization problems. In the simulation, we used the same number of agents as it was used in the original GWO. For GWO, this number is optimal to solve most optimization problems in good time. The open question remains how the results would change if we used an order of magnitude more agents in a 2D P colony and used massive parallelism.

Acknowledgments. This work was supported by The Ministry of Education, Youth and Sports from the National Programme of Sustainability (NPU II) project "IT4Innovations excellence in science - LQ1602".

Research was also supported by the SGS/11/2019 Project of the Silesian University in Opava.

References

1. Cienciala, L., Ciencialová, L., Langer, M.: Modelling of surface runoff using 2D P colonies. In: Alhazov, A., Cojocaru, S., Gheorghe, M., Rogozhin, Y., Rozenberg, G., Salomaa, A. (eds.) CMC 2013. LNCS, vol. 8340, pp. 101–116. Springer, Heidelberg (2014). https://doi.org/10.1007/978-3-642-54239-8_9

2. Cienciala, L., Ciencialová, L., Perdek, M.: 2D P colonies. In: Csuhaj-Varjú, E., Gheorghe, M., Rozenberg, G., Salomaa, A., Vaszil, G. (eds.) CMC 2012. LNCS, vol. 7762, pp. 161–172. Springer, Heidelberg (2013). https://doi.org/10.1007/978-3-642-36751-9_12
3. Ciencialová, L., Csuhaj-Varjú, E., Cienciala, L., Sosík, P.: P colonies: survey. J. Membr. Comput. **1**(3), 178–197 (2019)
4. Kelemen, J., Kelemenová, A., Păun, Gh.: Preview of P colonies: a biochemically inspired computing model. In: Workshop and Tutorial Proceedings. Ninth International Conference on the Simulation and Synthesis of Living Systems (Alife IX), Boston, Massachusetts, USA, pp. 82–86, 12–15 September 2004
5. Pavel, A.B., Buiu, C.: Using enzymatic numerical P systems for modeling mobile robot controllers. Nat. Comput. **11**, 387–393 (2012). https://doi.org/10.1007/s11047-011-9286-5
6. Păun, Gh: Computing with membranes. J. Comput. Syst. Sci. **61**(1), 108–143 (2000)
7. Păun, Gh, Păun, R.: Membrane computing and economics: numerical P systems. Fundam. Inform. **73**(1–2), 213–227 (2006)
8. Păun, G., Rozenberg, G., Salomaa, A.: The Oxford Handbook of Membrane Computing. Oxford University Press Inc., Oxford (2010)
9. Mirjalilia, S., Mirjalilib, S.M., Lewisa, A.: Grey wolf optimizer. Adv. Eng. Softw. **69**, 46–61 (2014)
10. Nishida, T.Y.: Membrane algorithms. In: Freund, R., Păun, G., Rozenberg, G., Salomaa, A. (eds.) WMC 2005. LNCS, vol. 3850, pp. 55–66. Springer, Heidelberg (2006). https://doi.org/10.1007/11603047_4
11. Nishida, T.Y.: Membrane algorithms: approximate algorithms for NP-complete optimization problems. In: Ciobanu G., Păun G., Pérez-Jiménez M.J. (eds) Applications of Membrane Computing. Natural Computing Series, pp. 303–314. Springer, Heidelberg (2006). https://doi.org/10.1007/3-540-29937-8_11
12. Valenta, D., Ciencialová, L., Langer, M., Cienciala, L.: Modelling of grey wolf optimization algorithm using 2D P colonies. In: ITAT 2020, Information technologies - Applications and Theory 2020, ITAT 2020 Conference Proceedings, CEUR Workshop Proceedings, vol. 2718, pp. 192–200 (2020)
13. Valenta, D., Langer, M.: On 2D P colonies and grey wolf algorithm. In: SGEM 2020, Sofia: 20th International Multidisciplinary Scientific GeoConference SGEM 2020, SGEM2020 Conference Proceedings, to appear, vol. 20, pp. 231–238 (2020)
14. Zaharie, D., Ciobanu, G.: Distributed Evolutionary Algorithms Inspired by Membranes in Solving Continuous Optimization Problems. In: Hoogeboom, H.J., Păun, G., Rozenberg, G., Salomaa, A. (eds.) WMC 2006. LNCS, vol. 4361, pp. 536–553. Springer, Heidelberg (2006). https://doi.org/10.1007/11963516_34
15. Zhang, G., Cheng, J., Gheorhe, M.: A membrane-inspired approximatealgorithm for traveling salesman problems. Rom. J. Inf. Sci. Technol. **14**, 3–19 (2011)

Author Index

Printed in the United States
by Baker & Taylor Publisher Services